"This book expanded my imagination about what is possible in groups of any form. This expansion came from the careful articulation of multidimensional, multi-evolutionary science and its union with functional behavioral science. The book provides a highly useful and operational approach for building and transforming groups to be more prosocial. This book is chock-full of clear steps on how to engage the Prosocial process to increase prosociality in the world. The book is a welcome, practical, and well-documented guide for people interested in lifting up the capability and competence of a group to be more productive, equitable, and collaborative."

—**Jane Dutton, PhD**, professor emerita of business administration and psychology, and coauthor of *Awakening Compassion at Work*

"This groundbreaking book shows how evolutionary theory can be used to empower the success of small groups. It is a seamless integration of the processes of psychological flexibility, the science of organizational development and interpersonal trust, and Elinor Ostrom's Nobel Prize–winning 'core design principles' for group cooperation. What results is a method that is both principle-based and highly flexible, and that speaks directly and practically to the role of coaches and consultants. This book is essential reading for anyone who works with people in group settings, and who is committed to helping these groups to be their best."

—**Susan David, PhD**, cofounder of the Institute of Coaching— a McLean/Harvard Medical School Affiliate; psychologist at Harvard Medical School; and author of *Emotional A*...

"What do you get when you combine contemporary evolutionary theory, insights from a winner of the Nobel Prize in Economics, and years of psychological research on behavior change? As it turns out, you get a practical and intuitively appealing set of steps for increasing care and cooperation amidst the twenty-first century's soul-sapping and ecologically damaging selfishness. The approach developed by Atkins, Wilson, and Hayes shows real promise, and I hope many people adopt it."

—**Tim Kasser, PhD**, professor in the department of psychology at Knox College in Galesburg, IL; and coauthor of *Hypercapitalism*

"A powerful, simple, but new perspective; part what to do, mostly how to do it, on the big question. It's always been make-or-break for our species. Somehow, it's looming ever larger: 'How can I be for myself, but also for others and our world?'"

—**Nicholas Gruen, PhD**, visiting professor at Kings College London Policy Institute, CEO of Lateral Economics, and author of the forthcoming *The Public Goods of the 21st Century*

"*Prosocial* helps feuding individuals figure out their common purpose, and gives warring factions the means to cooperate. Anyone who has ever tried to solve a burning conflict within or between groups should read this book."

—**Susan Pinker**, psychologist, behavioral science columnist for *The Wall Street Journal*, and author of *The Sexual Paradox* and *The Village Effect*

"One of the authors, David Sloan Wilson, has said that everyone needs to become 'wise managers' of their own cultural evolutionary process. This book is a great way to learn how to do this very thing. If you are looking for ways to achieve shared goals, work effectively in groups, or in any way make the world a better place, you are going to want to read this book. It is a deeply practical and insightful guide for individuals and the organizations they work for."

—**Joe Brewer**, executive director of the Center for Applied Cultural Evolution

"Evolutionary science has a lot to say about how we can create more effective, cooperative groups. So does the research of Nobel Laureate Elinor Ostrom, the great scholar of the commons. Blending these two rich streams, Atkins, Wilson, and Hayes provide powerful, practical guidance for anyone wishing to nourish prosocial behaviors—in education, business, civic action, athletics, and beyond. A concise, readable synthesis that practitioners and scholars alike will appreciate."

—**David Bollier**, coauthor of *Free, Fair, and Alive*

PROSOCIAL

USING EVOLUTIONARY SCIENCE

TO BUILD PRODUCTIVE, EQUITABLE,

AND COLLABORATIVE GROUPS

PAUL W.B. ATKINS, PhD
DAVID SLOAN WILSON, PhD
STEVEN C. HAYES, PhD

CONTEXT PRESS
An Imprint of New Harbinger Publications, Inc.

Publisher's Note

This publication is designed to provide accurate and authoritative information in regard to the subject matter covered. It is sold with the understanding that the publisher is not engaged in rendering psychological, financial, legal, or other professional services. If expert assistance or counseling is needed, the services of a competent professional should be sought.

In this book, some clients have been anonymized and descriptions of some groups are combinations of groups that have been modified for pedagogical reasons.

In consideration of evolving American English usage standards, and reflecting a commitment to equity for all genders, "they/them" is used in this book to denote singular persons.

Distributed in Canada by Raincoast Books

Copyright © 2019 by Paul W. B. Atkins, David Sloan Wilson, and Steven C. Hayes
                          Context Press
                          An imprint of New Harbinger Publications, Inc.
                          5674 Shattuck Avenue
                          Oakland, CA 94609
                          www.newharbinger.com

Figures 1.1 and 1.2 reproduced courtesy of William Muir.

Cover design by Amy Shoup; Acquired by Catharine Meyers;
Edited by James Lainsbury; Indexed by James Minkin

All Rights Reserved

Library of Congress Cataloging-in-Publication Data on file

Paul Atkins: To Betty, my mother, who taught me about being prosocial. I wish I could have shown you this book.

And to Khia, Dana, and Evie, whose love and support throughout the writing of this book sustained me.

David Sloan Wilson: To my nuclear family, Anne, Katie, and Toby, a prosocial unit if ever there was one.

Steven C. Hayes: To Steven J. Pistorello-Hayes, my sweet "little boy" who is not so little anymore. I hope, in some small way, that what we have done with the Prosocial process will touch the lives of children, who so deserve a more prosocial and nurturing world in which to live.

Contents

Foreword		vii
Introduction		1

Part 1: Concepts and Principles

Chapter 1	Evolution at Multiple Levels and in Multiple Streams	10
Chapter 2	Elinor Ostrom and the Commons	22
Chapter 3	Core Design Principles, Version 2.0	35
Chapter 4	Evolving Behavior *A Contextual Behavioral Approach*	49

Part 2: Prosocial Methods

Chapter 5	Mapping Interests and Building Psychological Flexibility with the ACT Matrix	72
Chapter 6	Modules and Pathways for the Prosocial Process	95
Chapter 7	Core Design Principle 1 *Shared Identity and Purpose*	102
Chapter 8	Core Design Principle 2 *Equitable Distribution of Contributions and Benefits*	117
Chapter 9	Core Design Principle 3 *Fair and Inclusive Decision Making*	135
Chapter 10	Core Design Principle 4 *Monitoring Agreed Behaviors*	151
Chapter 11	Core Design Principle 5 *Graduated Responding to Helpful and Unhelpful Behavior*	162
Chapter 12	Core Design Principle 6 *Fast and Fair Conflict Resolution*	176

Chapter 13	Core Design Principle 7 *Authority to Self-Govern*	189
Chapter 14	Core Design Principle 8 *Collaborative Relations with Other Groups*	199
Chapter 15	Goal Setting for Action	209
Conclusion	The More Beautiful World That Our Hearts Know Is Possible	218
Endnotes		227
References		237
Index		251

Foreword

As we look to the world's future, we see formidable problems to be solved. We are rapidly polluting our planet, overheating our atmosphere, and siphoning more than half the wealth and resources of our entire planet into the hands of a few dozen individuals. We engage in competitive relations within and between people and organizations that are not only wasteful, but a source of suffering to many. Dire poverty sits nakedly adjacent to excessive consumption and conspicuous greed, a dynamic that too often erupts into cycles of violence and oppression.

Many people claim these more destructive trends in modernity are manifestations of human nature. But to what human nature are they referring? Clearly, there are many sides to humans, and the "natures" they express. People are also helpful, friendly, and cooperative. Everywhere we can find people working together in small groups and communities for the common good. Daily, in virtually every society across the globe, we can witness love, collective sacrifice, and caring for others.

It is with these multiple human natures in mind that I find this book by Atkins, Wilson, and Hayes to be both so timely and so important. Harnessing the Nobel prize–winning work of Elinor Ostrom, these authors build on both evolutionary theory and cutting-edge behavioral science to specify the characteristics of groups that can not only do well by their members but also forge a positive direction in relation to the rest of the ecologies in which they are embedded. Underlying the deep theory informing Prosocial practices is an understanding of our complexly evolved human natures. Yes, we can be competitive and destructive. But we can also be cooperative, constructive and giving. Which human nature we display is not just a matter of genetics, but also of the social-environmental designs in which we find ourselves. Understanding the core elements of such environments is a key to catalyzing the best of our evolved human propensities and capabilities—which, as it turns out, are manifold!

Accordingly, *Prosocial* is not just a naïve manifesto that describes "good" organizations and groups. Instead, Atkins, Wilson, and Hayes apply their collective expertise as scientists, clinicians, and consultants to not only describe an idealized vision of how groups can function, but also to provide concrete plans for arriving there. Across these chapters the authors lay out specific principles and practices, many drawn from their frontline experiences as change agents, that, when embraced, can enhance the common welfare of stakeholders. These practical steps focus on topics from communication to equity within organizations—detailing the elements that can foster cohesion, effectiveness, and stability.

It's hard to imagine any organizational leader or engaged member who would not find some ideas of merit in these pages. Concretely, there is some great advice on issues such as goal setting, organizational structures, conflict resolution, and motivation that anyone can use. Yet perhaps more important, the authors show convincingly that, when people are working well together in accord with the core Prosocial principles they articulate, they end up with goals they can all endorse and stand behind, and thus they work harder and better *together* for their organizations. That is, having a cohesive, shared value–driven group pays off in terms of effectiveness, productivity, and satisfaction, all contributing to the bottom-line goals of organizations.

The emphasis in Prosocial is thus primarily on enhancing group functioning. Without doubt, some of our contemporary problems could be ameliorated by changing the minds and actions of individuals. People could consume less, use less energy, and be more generous to and tolerant of others. But it seems unlikely that individual motivations to change will be sufficient, even when aggregated. So many of the issues we face are fostered by the aims, actions, and dynamics of organizations, identity groups, and nation states, rather than single persons, and it is these groups who will be needed to create positive change. Corporations, educational institutions, nonprofit organizations, and advocacy groups will need to begin to care not only about the welfare of their own stakeholders, but also about their groups' relations to the other groups with which they cooperate or compete. Groups, working with purpose and cohesiveness, will be needed to mobilize collective action and reforms.

Yet for this reader, another aspect of the work was especially resonant—its support for the autonomy, competence, and relatedness of both individuals and the groups with which they identify. In my own work within self-determination theory, I have spent more than 40 years researching what motivates people to make sustainable changes in behavior and what conditions are essential to their wellness and thriving. Across that work it has been clear that the capacity to be autonomous—to endorse one's own actions—is critical. Prosocial emphasizes the importance of autonomy in its core design principle 3 (fair and inclusive decision making) as well as the collective autonomy of groups in its core design principle 7 (authority to self-govern). It thus helps groups, both large and small, to find their own inner compasses, and then to chart their own destiny. Prosocial also highlights the limitations of relying upon coercive regulations to achieve group goals. Finally, it supplies concrete strategies for developing the kind of cohesive functioning that can enhance a sense of belonging and effectiveness within groups and organizations as they pursue those aims that matter most to them.

An implicit but continuous message in this book is that of responsibility: that we each have our parts to play in changing our groups, companies, and governments for the better. As we look ahead to the 21st century challenges we collectively face, it is indeed no time for passivity or idle hope. We need to actively change how our groups, organizations and societies are functioning *now*. *Prosocial* is thus appropriately also a call to action, as well as being a tutorial on what kinds of actions are likely to be most effective.

Because it is individuals who are needed now to inspire their groups to move in the right direction, I am thus happy to recommend that you read this book. Share it with your coworkers, your leaders, and your communities. Then, together, move ahead with the inspirations it provides. Through collective action, we can meaningfully enhance the situation our children and grandchildren will inherit from us. This little planet Earth is, after all, the only commons we have to share with them. If the authors of this work are right, taking these actionable steps will not only provide some long-term promise for the coming generations, but also will be intrinsically rewarding to all who take up the challenge.

—Richard M. Ryan
Sydney, NSW, AUS, and Rochester, NY, USA

Introduction

In May 2014, the deadly Ebola virus spread from the African nations of Guinea to Sierra Leone. Within months, first hundreds and then thousands were dying. Few villagers understood the causes of the disease, and after years of war and strife some so deeply mistrusted outsiders that they blamed the aid workers for the horror itself. The BBC reported that in Guinea eight health workers and journalists were murdered. The rapid spread of the disease was exacerbated by burial practices that included the washing and kissing of the dead as a sign of grief, respect, and love. Against this backdrop, the attempts of aid workers dressed in strange "spacesuits" to stop such practices, enforce strict quarantines, and isolate the dead in hazmat body bags seemed destined to fail.

But in the Bo District of Sierra Leone, a group of committed volunteers tried a different approach. They worked with the villagers themselves to find ways of responding that addressed the urgent need to stop the spread of the virus while also respecting local customs and values. Instead of imposing solutions from the outside, the villagers talked about what mattered to them—cherishing their dead and, at the same time, stopping the spread of disease—and how best to accomplish those aims in ways that made sense to them. One group of villagers came up with a plausible solution. Instead of keeping a loved one's body at home, they chose to substitute a banana trunk for the body. In this way, all the customary rituals, including washing, dressing, and kissing, could be performed on the symbolic body before the body was buried. Relatives who adopted this approach felt satisfied that they could express their love and respect, without further endangering their community. With the help of the volunteers, the new practice spread rapidly through the Bo District, helping to bring the disease under control.

Around the same time, the chief executive officer of the Museum of Australian Democracy, one of the main cultural institutions in Canberra, was wondering how she could unite her team around a common purpose and better resolve internal conflicts and disagreements that had held the museum back from its central role of educating the public and preserving cultural artifacts. She turned to a skilled team of facilitators from the Australian National University, and together they implemented a series of team meetings and individual coaching sessions that built a sense of trust and safety within the organization. Employees began to openly discuss the problems and challenges they saw, while also working together to come up with a strategic direction for the future, a set of actionable goals related to this direction, and a concrete action plan for achieving those goals. Because the intervention occurred between two staff opinion surveys of the entire Australian Public Service, it was possible to see how these interventions affected museum staff relative to all other government agencies. The results were astonishing. In a year when the morale of the average Australian civil servant dropped dramatically, the museum staff's opinions regarding the quality of management, the value of their work, and their level of job satisfaction improved greatly.

A few years before these two examples unfolded, the superintendent of Binghamton, New York's public schools asked one of the authors of this book (David) to advise on the creation of a "school within a school" for at-risk youth. David designed the school using eight core design principles that all groups need to function well, plus two additional auxiliary design principles especially relevant to education. To enter this Regents Academy, as it was called, ninth- and tenth-grade students had to have flunked at least three of their courses during the previous year. Within the first quarter at the Regents Academy, these students were performing remarkably better than a comparison group in a randomized controlled trial—the gold standard of scientific assessment. By the end of the year, they were performing as well on the state-mandated exams taken by all students as the average high school student in the Binghamton public school system. By this measure, at least, the learning deficits of previous years had been erased.

In these three examples you can see enormous differences in the size of the groups mentioned; in the resources at their disposal; and in their cultures, values, and aims. For example, improving educational outcomes was the aim with the United States example, improving services to citizens with the Australia example, and simply stopping more people from dying with the Sierra Leone example. So what do they have in common? In each example a group of individuals, operating in a given context, was trying to achieve a common goal. The members of these groups all faced the same challenge, which can be summed up with this question: How can we support ourselves in order to act together in such a way that satisfies our own interests, the interests of our group, and the interests of the larger systems of which our group is a part? This book is an attempt to answer that question.

These groups also had another thing in common. They were all helped by the Prosocial process described in this book, which helps groups, and groups of groups, create more strongly shared identity, greater engagement in purpose, better relationships, and improved performance. We will soon dive into the topic of how your group can benefit from the Prosocial process, but first we want to establish the wider stage upon which this work rests: By improving your group, you are literally doing the work of intentional cultural evolution. You are creating a more cooperative world.

In an age that recently was dubbed the Anthropocene, a name that reflects our impact on planet Earth, we are increasingly faced with the prospect of either global cooperation or ruin. Technology has truly created a global village, in which talking to someone halfway around the world is as easy as talking to our next-door neighbor. With a few clicks of a button, we can schedule to stay in a person's house on the other side of the planet, share a lawnmower, learn a specialist skill, or join a global movement for change. But as much as the Internet has increased the potential for cooperation, it is also true that our most challenging problems still all have roots in our failure to cooperate. Global issues, including climate change, habitat and species loss, global financial crises, income inequality, trade wars, and the effects of technology upon relationships, and the seemingly individual issues, such as escalating rates of obesity and depression, all have strong social elements embedded within them.

Alexis de Tocqueville, the great French social theorist who visited America in the 1830s, observed that "The village or township is the only association which is so perfectly natural that, wherever a number of men are collected, it seems to constitute itself."[1] Tocqueville recognized that small groups seemed to naturally work well. As we explore in chapter 1, for most of our history as a species we have evolved in the context of small cooperative groups, so we know how to do small groups very, very well, at least compared to every other species on the planet.

But compared to even the 1830s, when Tocqueville was writing, the world is now a very different place, and social interactions with strangers at larger scales, far beyond those of the tribe or village, create challenges for humans. These days, the value of the small group has been almost eclipsed either by a focus on the individual, or on vast groups at a national or even global scale. In the 1960s, the "small world" experiment by Stanley Milgram suggested that any two people in the US could connect with one another through approximately six people. This became the famous idea of six degrees of separation.[2] In 2016, Facebook was able to give a much more precise estimate, showing that any two individuals in its social network of 1.59 billion users were connected by only 3.46 people, and this separation is decreasing all the time.[3]

We are, in effect, in a state of evolutionary mismatch. We evolved to be extraordinarily cooperative in small groups, but we are now faced with the challenge of a wholly new environment in which we're asked to cooperate, not just with those we know well but also with people who are practically strangers—that is, not just with "us" but also with "them." When our social environments become different than the contexts in which we evolved, then what used to be "natural" and "effortless" can go terribly wrong— even in small village-sized groups. We are in entirely new territory today. Never before have we so frequently been faced with the challenges of coordinating with groups with perspectives that differ radically from that of our own. But, also, never before have so many people had access to the tools to create cooperation on a vast scale.

In this book we describe core design principles for building better-functioning groups, based on the Nobel Prize–winning work of Elinor Ostrom, as well as behavior change techniques that also enhance group functioning. We call this the Prosocial process, which works on three main fronts and is focused on building better small groups, or groups of small groups, in this new evolutionary environment humans are creating:

- First, the Prosocial process helps groups build a sense of "us" beyond each member's individual self-interest and determine the group's reasons for existing. No group can perform well unless its members have a clear sense of shared purpose and identity.

- Second, this process improves cooperation within groups by helping members balance individual self-interest and the interests of the collective through the application of the core design principles. The Prosocial process strives to build upon the passion and motivation that arises from acting in line with one's own needs, values, and aims, while also satisfying the needs, values, and aims of others in the group, and the collective as a whole.

- Third, and perhaps most importantly, the principles and the behavior change techniques of the Prosocial process help individuals coordinate between groups, allowing for the possibility that we as humans might create more equitable, effective, and satisfying systems of cooperation regionally, nationally, and globally.

This book is written for anyone who wishes to improve cooperation within and between groups. No special background is required. You might be a university professor, a policy expert, a politician, a pastor, a teacher, a therapist, the CEO of a business or nonprofit, a community activist, a parent, or even a high school student. If you care about increasing cooperation, then our hope is that this book will give you the framework you need to do so in whatever groups you care about.

Defining Prosocial

The word "prosocial" and its synonyms and antonyms occur repeatedly throughout this book, so let's get clear on what we mean. *Prosocial behavior* is generally understood as the act of getting along and cooperating with others. It is about benefiting others or society as a whole. Prosocial behavior includes *altruism*, which involves self-sacrifice, but we don't limit our definition to helping others at a cost to oneself. Indeed, we are more interested in creating win-win situations in which everyone gains.

What is the opposite of prosocial behavior? Uncooperative behaviors can be either intentional or unintentional. In this book we use words such as "selfish" to describe intentional acts designed to optimize outcomes for the self over others, and words such as "antisocial" to mean a willful intent to harm others. But much of the time failures to cooperate don't really involve much consideration of others, so we use words such as "self-interested," "uncooperative," or "unhelpful" to refer to behavior that privileges self-interest over collective interests but isn't necessarily intended to harm the collective.

The word "prosocial" is derived from the Latin *pro*, meaning "in favor of," and *socialis*, meaning "allies or friends." But in an increasingly interconnected world, we can no longer afford to just be in favor of our friends. For example, members of terrorist groups may be caring and considerate of one another while killing others with different beliefs and practices. Joshua Greene, in his outstanding book *Moral Tribes*, distinguishes between the "me versus us" problem of cooperation within small groups and the "us versus them" problem of cooperation between groups. He argues that the "us versus them" problem is much harder to address because we lack evolutionarily evolved responses to get along well with strangers.[4] While we agree with this point, in this book we wish to develop it in two main ways: first, by recognizing that there is also a "me versus me" problem when we seek to disown, deny, or avoid parts of ourselves (see chapters 4 and 5), and, second, by offering a practical approach to improving group cooperation between groups (not just within them) that can soften "us versus them" by nesting both in a larger "us."

One of the challenges of writing about prosociality is that it is a relative concept, and not something absolute. What is "helpful" or "cooperative" behavior ultimately depends upon your point of view. Is it more cooperative to tactfully hold your tongue or boldly tell your truth? One course of action will provide social support while another might provide helpful information. Or, consider a soldier who helps his comrades kill the enemy. If you are a comrade, this behavior is as prosocial as it gets—risking one's life to support one's allies. But, of course, the enemy experiences it as the height of antisocial behavior. You can see that prosociality is, in this respect, the same as morality. It depends on the eye of the beholder.

But this relativity does *not* mean that we, the authors, are relativists who regard all social constructions as equal. Some values and goals are "objectively" better than others in the sense that they lead to greater human thriving. Human thriving, then, is our benchmark, and thriving involves both agency and communion. As humans, we do better when we have the competence and self-determination to act effectively in the world in ways that align with our values and goals. And people who value high-quality relationships over individual wealth, fame, or power are happier than those who are motivated by more extrinsic goals and values.[5] None of us are more or less important than anyone else, and ultimately our work is grounded in an ethic of caring for the well-being of others and respecting individual self-determination in the context of collective good. In the end, "prosocial" will always be a contested term, as are "leadership," "justice," and "morality." What's important is paying attention to working to increase prosocial behavior.

How This Book Is Organized

Our aims in this book are to:

- educate you on the Prosocial process,

- clarify why humans interact in groups the way they do, and

- give you the tools and ideas you need to start implementing the Prosocial process in groups and groups of groups of your own.

This book is part popular science book and part how-to. In it we'll explain important concepts *and* show you how to implement those concepts. A core idea of our approach, as you will see, is that the locally appropriate, contextually relevant solutions that groups grow for themselves are usually much more effective than trying to shoehorn in the latest fad or conceptual model. Our job as we see it is to direct attention toward the key principles that groups need to understand and use in order to function more effectively, and to give groups the tools they need to have deep and rich conversations about how best to implement those principles in ways that will support the group (or group of groups) to achieve its purposes. As you will see, this book is as much about helping groups of groups to cooperate as it is about helping groups of individuals.

We've organized this book into two parts. In the first part we explore how and why evolutionary theory and Elinor Ostrom's Nobel Prize–winning work on the commons provide the basis for the Prosocial process. Chapter 1 describes how human cooperation can be understood as a complex adaptive and multilevel system that is evolving. It is hard to plan, predict, and control when you're dealing with a complex adaptive system, but evolutionary theory can help you chart a path toward sensing and responding consciously in order to move small and large groups in directions you wish to go.

In chapter 2 we explore how Ostrom's work to develop the core design principles associated with groups effectively sharing resources can be broadened to improving cooperation in all groups. We also explore the pernicious effects of believing humanity is basically selfish, as well as the idea of the "commons." We see the latter as an aspirational vision for how we could organize locally and globally, making use of markets and regulation while steering a course between excessive regulation and systems that are unhooked from basic human values, such as fairness, loyalty, and caring for each other and the planet.

In chapter 3 we describe how we broadened Ostrom's original core design principles to make them not just descriptive of good practice but also prescriptive, and not just applicable to groups that share resources but to all groups or groups of groups.

In chapter 4 we examine the social psychology of cooperation and conclude that groups need to employ methods that promote trust, future thinking, and a social values orientation. After reviewing traditional learning principles and the science of symbolic learning, we describe how these principles undergird the need for psychological flexibility and show that its elements are both consistent with evolutionary thinking and also useful for improving relationships between the parts of ourselves, between ourselves and others, and between the groups we form.

In part 2 of the book we turn our attention to describing in more detail the tools you will need to help small groups function more effectively. In chapter 5 we explore the acceptance and commitment training matrix, a tool that teaches people to pay attention to two primary dimensions of experience: (1) the "inner" and "outer" aspects of experience and (2) the behaviors that move them toward what matters (including prosocial relations with others) or away from fear and other difficult emotions. The matrix is relatively simple to learn, is useful for promoting psychological flexibility, and can be scaled to any level, from the individual to groups of groups.

In chapter 6 we present an overview of the modules that make up the Prosocial process, as well as some ideas about how they can be ordered in different circumstances as variants of the process.

Then in chapters 7 through 14 we explore each of the eight core design principles. For each principle we briefly review the science of why it's important for enhancing cooperation, then we explore ways to build healthier, more flexible conversations within groups to develop appropriate and effective agreements to instantiate the principle.

Although we treat each principle separately, they are, in fact, integrated, and wherever possible we point out how the principles are interwoven with one another.

In chapter 15 we explore ways to set goals with a group. No matter how much awareness a group conversation may create, the awareness is worthless unless it's converted into real actions that create change. Goal setting helps groups achieve this. However, though this step is relatively straightforward, it is often forgotten.

Finally, we conclude by exploring the wider implications of the Prosocial approach. Although we see our efforts in this book as just the beginning of a process that will ultimately be carried forward by a global community, we can already see how this approach can easily be scaled to address regional, national, and even global challenges.

Before starting this journey, and especially because the part 1 chapters are necessarily more conceptual than practical, it might help to consider what you want to improve in your group. If you could wave a magic wand and improve any aspect of your group's functioning, what would it be and why? Mentally stick that thought in your back pocket and carry it with you as you read the book, treating it as a kind of touchstone, and see if this book can step up to address the implicit and explicit needs of your group—the reason you picked this book up in the first place.

PART 1

Concepts and Principles

CHAPTER 1

Evolution at Multiple Levels and in Multiple Streams

As humans, we have a mixed record with intentionally bringing about change. While our capacity to change the physical world seems almost limitless, our capacity to manage behavior and bring about desired cultural change appears to have developed little over millennia. This is true for us as individuals—just look at how hard it is to change diet and exercise regimes—but also for the groups in which we participate.

We think we can do better, if we let science guide us.

In this chapter we turn to the evidence from evolutionary science that not only tells us where we come from but also provides hints about what we need to do to consciously shape our behavior in order to be more prosocial. Though you might be used to thinking about evolution in terms of biology, evolutionary principles can just as easily be applied to thinking about changes in behavior, language, and culture. Indeed, we believe that evolution must be at the center of thinking about behavioral and cultural change in order for us to step up to the challenges the modern world gives us. Unless we become "wise managers of evolutionary processes," social evolution might well take us where we do not want to go.[1]

This chapter will introduce you to some key evolutionary principles. They are relatively simple to understand but combine to form a powerful way of comprehending change at multiple levels of organization. We believe it's necessary to broadly understand the fundamentals of evolution if we are to tackle the extremely tough challenge of increasing cooperation, particularly between groups of groups. Furthermore, as you'll see in chapter 4, the very same principles that underlie biological evolution are also useful for understanding symbolic and cultural evolution, so your time spent learning about them will be doubly rewarded. The principles can be scaled to nested networks of groups, allowing humanity to take better charge of our social evolution to create more equitable, harmonious, and purposeful relations for us all.

That may sound like a tall order, but let's begin at the beginning and see.

Darwin and the Fundamental Problem of Social Life

The dominant cultural worldview that provided the background for the development of Charles Darwin's theory of natural selection assumed that God's creation was well designed from top to bottom, from the stars in heaven to the tiniest insect. The triumph of Darwin's theory lay in showing that design in nature can emerge from a purely natural process based on "the laws acting around us,"[2] as he put it in the final passage of *On the Origin of Species*. Those laws prevail when three conditions are met: (1) when individuals show *variation* in their measurable properties, (2) when those variations make a difference in terms of survival and reproduction in a given environment (collectively called *fitness*, or *selection* if you wish a more general term), and (3) when offspring resemble parents (collectively called *heritability*, or *retention* of traits). Under these conditions of variation, selection, and retention, the measurable properties of individuals in a species change over the course of multiple generations. Traits that make an organism more likely to survive and reproduce are selected for, and individuals become adapted to their environments as a result.

While natural selection is the centerpiece of Darwin's theory of evolution, not all products of evolution are adaptive. A given trait might arise by chance, might be a nonadaptive by-product of another trait, or might have been adaptive in the past but not in the present. Thus, the modern theory of evolution that is built upon Darwin's theory is multifaceted, and a lot of inquiry is often required to determine how a given trait evolved on a case-by-case basis.[3]

Although evolution is usually thought of as a lengthy process, it can occur rapidly enough to be seen. One of the earliest observed examples of visible natural selection in response to a changing environment involved the peppered moth in England. Prior to the onset of the industrial age, peppered moths hid against light-colored trees. As industrialization led to trees being covered in soot, it became increasingly easy for birds to detect and eat the light-colored moths, whereas moths that were relatively dark enjoyed a selective advantage. Since coloration is a trait that can be retained across generations, the population shifted to include more darker-colored moths. At no point was there an intention for the moths to become darker, or some mysterious controller. The coloration of the moths naturally varied, and the darker color was selected for by bird predation and then retained through genetics operating in a changing context of darkening trees.

At first, Darwin thought that natural selection could explain all examples of design in nature that had previously been attributed to a creator. However, he quickly encountered difficulties explaining social behaviors that are morally praiseworthy in human terms, such as honesty, bravery, and charity. These traits benefit others but are often costly for the individual. If natural selection favors individuals who survive and reproduce better than other individuals, then how can it favor individuals who enhance the survival and reproduction of others at their own expense? Wouldn't natural selection always reward selfishness? This problem can be seen as the fundamental problem of social life. Natural selection operating within a social group should tend to *undermine* the behaviors required for the group to function as a cooperative unit.

Darwin, however, perceived the outline of a solution to this fundamental problem to his theory. Yes, he reasoned, morally praiseworthy individuals might be at a selective disadvantage compared to more self-centered individuals within a social group, but *groups* of individuals who behave morally toward each other would have a decisive advantage over groups of self-centered individuals who were undercutting each other. As

Darwin put it: "Although a high standard of morality gives but a slight or no advantage to each individual man and his children over the other men of the same tribe...an advancement in the standard of morality will certainly give an immense advantage to one tribe over another."[4] Thanks to a process of between-group selection, Darwin saw that his theory could explain the evolution of cooperative traits expressed within groups after all—at least if certain conditions were met.

Between-group competition might take the form of direct conflict, as when a more cooperative group beats a less cooperative group in warfare, or it could be indirect, as when a more cooperative group survives a natural disaster better than a less cooperative group. Darwin was silent on the fact that group selection can only explain the evolution of cooperative behaviors within a group; it doesn't eliminate conflict so much as elevate it to the level of between-group interactions, where it can potentially take place with even more destructive force than before. To solve that problem, we need to add additional layers to the process of natural selection, as we shall see.

Multilevel Selection Theory

Fast-forwarding to the present, Darwin's insight has grown into a sophisticated and empirically supported approach to evolution called multilevel selection (MLS) theory, which we'll now briefly explore. Our aim is to help you see the world through the lens of continuous and simultaneous evolution that's occurring at multiple levels. Once you see life in that way, you'll be much better equipped to understand how the principles we introduce later can enhance group functioning at any level.

MLS theory can be summarized in terms of the following tenets:

1. Social behaviors are almost invariably expressed among sets of individuals (groups) that are small compared to the total evolving population. Thus, almost all evolving populations are *multigroup* populations. One religious group may be prospering, another shrinking; one business is growing, another is going bankrupt. The groups form, dissolve, and are connected to each other in diverse ways in different populations.

2. As a basic matter of tradeoffs, behaviors that are for the good of the group often have costs for the individual doing the behaving, and therefore they seldom improve the relative fitness of the individual within the group. Even the smallest acts of kindness, such as opening a door for someone, typically require at least a small amount of time, energy, and risk on the part of the prosocial individual. These costs are amplified with more substantial acts, such as helping others develop skills, creating social events, or consoling someone who is upset. This makes natural selection at the smallest scale—among individuals within a single group—primarily *disruptive* as far as prosociality is concerned.

3. Because of the first two tenets, between-group selection—a process by which certain groups prove themselves fitter than others in the context of their

environments—is typically required in order for prosocial behaviors to evolve. Furthermore, between-group selection must be strong enough to oppose the disruptive effects of within-group selection. Said another way, cooperation and prosociality are selected for because they benefit the group that's competing with other groups more than they hurt the individual who's competing with other individuals within the same group. Multilevel selection is like a tug-of-war, with within-group selection pulling toward self-serving traits and between-group selection pulling toward group-serving traits. This is why we see a mix of prosocial and disruptive self-serving behaviors in human life and why—with sufficient know-how—we can tilt the playing field in the direction of prosocial behaviors.

4. What we have shown for two levels of selection—between individuals within groups and between groups within multigroup populations—can be extended to all levels of selection. Extending the hierarchy downward, a single multicellular organism is a group of cells and genes. Extending the hierarchy upward, single species are embedded in multispecies ecosystems in the biological world. At every rung of this multitier hierarchy, the logic of within- and between-group selection repeats itself. In human terms, what's good for you can be bad for your family. What's good for your family can be bad for your clan. What's good for your clan can be bad for your nation. What's good for your nation can be bad for the world. The general rule is this: *adaptation at any level of a multitier hierarchy requires a process of selection at that level—and it tends to be undermined by selection at lower levels*. Or, as David, one of this book's authors, concluded in an article written with Harvard biologist Edward O. Wilson, "Selfishness beats altruism within groups. Altruistic groups beat selfish groups. Everything else is commentary."[5]

Before we explore what all this means for contemporary human groups, let's consider two examples from the natural world that nicely illustrate how multilevel selection works. First, consider cancer, which is really a process of natural selection among cells taking place within a single multicellular organism. Mutations occur with every cell division; some mutations cause the mutant cells to proliferate faster than the neighboring cells. With additional mutations, some groups of mutant cells become tumors that interfere with bodily function. *Metastasis* is the evolution of the ability of mutant cells to disperse to other parts of the body. All of this counts as adaptive from the perspective of the cancerous cell. The fact that cancers bring about their own death with the death of the organism is irrelevant. Evolution has no foresight and natural selection within groups (in this case a multicellular organism) is insensitive to the welfare of the group.

The number of cancer-causing mutations is thankfully rare compared to the total number of cell divisions, and this is due to several hundred million years of natural selection at the level of multicellular organisms that have cellular differentiation. Very simply, cancer-resistant organisms survived and reproduced better than cancer-susceptible organisms. Scientists are only beginning to understand the arsenal of cancer-preventing mechanisms that operate in our bodies, as naturally as our vision, our breath, our beating hearts, and our ability to cooperate in small groups.

Now let's move from cells within organisms to organisms within groups. Imagine that you're a chicken breeder who wants to increase the egg-laying productivity of hens. In commercial egg farms hens are raised in cages—in small groups. If you want to increase egg production, should you allow the most productive hens to breed, or all of the hens in the most productive cages (even if some hens in that cage do not lay many eggs)? An animal breeder named William Muir at Purdue University actually performed an experiment to answer this question, which led to a somewhat surprising outcome—surprising, that is, if you don't understand MLS.[6]

When only the most productive hens were selected to breed, egg productivity declined over the course of five generations. Why? The most productive hens in each group achieved their productivity by bullying the other hens, a heritable behavior in chickens. And five generations of selecting for it, even if inadvertently, was sufficient to breed a strain of chicken psychopaths who plucked out each other's feathers and murdered each other (see figure 1.1). No wonder egg productivity declined!

Figure 1.1

When the most productive *groups* were allowed to breed, a strain of mild-mannered hens resulted, complete with an impressive increase in egg productivity over the course of only five generations. In this latter case, the experimenters strengthened between-group competition and reduced within-group competition by selecting at the cage, or small-group, level. As a result, more cooperative behaviors emerged. The birds kept their feathers and lives and peacefully laid their eggs (see figure 1.2). Needless to say, the poultry industry rapidly adopted the second procedure, so the eggs on your breakfast table come from strains of group-selected hens.

Figure 1.2

Though genetic evolution and examples such as cancer or inadvertently breeding selfish chickens might seem far removed from the topic of human cooperation, the concepts behind them do apply. Humans, too, are faced with the same choice that we observe in the natural world, between individual self-interest and the interests of the collective. Individual human behavior can foster success at one level at the cost of success at another.

Evolution at Multiple Levels

Once you get the hang of MLS theory, you can see processes of selection operating at multiple levels all around us. You'll see that a very large fraction of our problems are like cancer—adaptive at the local level in the evolutionary sense of the word but disruptive with respect to the higher-level common good. For example, leaders make decisions that benefit themselves at the cost of the organization, or they make decisions that benefit the organization at the expense of the broader community. We all make use of technologies that benefit each of us individually but collectively increase carbon dioxide emissions and exacerbate climate change. We see everywhere examples of self-interest at one level working against the collective interests at higher levels.

Paul told the story about the selective breeding of chickens to his university department, and it became a running joke. His colleagues immediately recognized themselves in the example, because they are often selected for promotion on the basis of how many publications they have individually, not on the basis of how helpful they are to the success of the whole group. Unfortunately, this means that an academic department can end up looking a little like the battered and bloody chicken coop of figure 1.1, in which chickens were selected for their individual egg laying and not for their capacity to get along with others!

We see the same patterns even in controlled studies. A fascinating example was conducted with American college students.[7] Incoming freshmen took a survey that measured their propensity to help others. Several months later, they took the same survey and were provided with three additional copies to give to friends that they had made in college. Then these groups (the focal individual and his or her three friends) played a game in which it was possible to win movie tickets for the whole group (a prosocial behavior) or for oneself (a selfish behavior). What do you imagine the results were?

Within each group, the more self-serving individuals (as measured by the survey) won more movie tickets than the more prosocial individuals. But that is not the whole story. The more prosocial individuals had found more prosocial friends in their first months in college. As a result, the more prosocial groups collectively earned more tickets than the more selfish groups. The two levels of selection (self-serving and prosocial) roughly canceled each other out. There was no overall relationship between individual prosociality and the number of movie tickets won because the between-group advantage of prosociality counterbalanced the within-group advantage of selfishness.

Obviously, genetic evolution did not take place in this experiment, but its results showcase the same fundamental principles found in the nonhuman examples we discussed. Every one of us, in every group that we take part in, has the option of maximizing our personal advantage compared to other members of our group, or maximizing our collective advantage as a group. As a basic matter of tradeoffs, these two options usually require different behaviors. What it takes to grab the biggest slice of the pie is different than what it takes to make the biggest pie. This is, quite literally, the fundamental problem of human social life, and it is just as true for an individual in a team as it is for a nation-state arguing that it should be allowed to release more carbon than other nations.

Evolution in Multiple Streams

So far we have expanded the traditional view of evolution by pointing out that selection can occur within and between multiple levels simultaneously. But we need to go further. To fully apply evolutionary thinking to human affairs, we need to introduce multiple systems of inheritance.

Darwin knew nothing about genes. For him, heredity meant a resemblance between parents and offspring, a fact that could be well documented without knowing anything about the underlying mechanisms. The discovery of genes as a mechanism of heredity, starting with Mendel's famous experiments on peas, was therefore a major breakthrough in evolution science. However, something important in our understanding of evolution was lost along the way. People quickly regarded genes as the *only* mechanism of inheritance, as if the only reason offspring resemble their parents is because they share genes. The textbook definition of evolution became "a change in gene frequency." Around the world and to this day, for experts and the lay public alike, when you say the word "evolution," almost everyone hears the word "genes."

Yet, only a small amount of reflection reveals this limited view of evolution to be patently false. Parents and offspring typically share the same language, for example, which is not genetically inherited. More generally, parents and offspring share a vast storehouse of learned information that is not genetically inherited. Yet, the concept of cultural evolution was almost totally ignored by evolutionary biologists until the 1970s. To make matters worse, other academic disciplines devoted to the study of human culture, such as anthropology, sociology, history, and linguistics, developed largely independently and sometimes in perceived opposition to evolutionary theory.

Thankfully, evolutionary scientists are going back to the basics by defining evolution in terms of variation, selection, and retention and not just genes. Academics and researchers recognize at least four mechanisms for the transmission of information across generations. The first is the well-known genetic inheritance system; offspring resemble parents because they share the same genes. The second mechanism is epigenetics, the shared expression of genes. While our interest here is not to describe epigenetic evolution, we do want to note that it occurs through the variation, selection, and retention of the mechanisms of expression, not the genes themselves—that is, the exact same evolutionary processes but applied to a different level of organization.

The third mechanism of inheritance is learning, which can be found in many species. The psychologist B. F. Skinner famously observed that trial-and-error learning is a variation-and-selection process much like genetic evolution. Consider what happens when a rat is placed in a "Skinner box" containing a lever that delivers a food reward when pressed. The rat explores its surroundings and eventually presses the lever by chance—much like a random mutation. Pressing the lever, unlike the other exploratory behaviors, results in a fitness-enhancing event (food). The rat has evolved, by genetic evolution, to repeat behaviors that result in fitness-enhancing events (experienced as pleasure) and avoid behaviors that result in fitness-reducing events (experienced as pain). It therefore selects the behavior of pressing the lever in the environment of the Skinner box to enhance its survival, similar to how a beneficial gene evolves over multiple generations when it enhances survival and reproduction. Skinner's phrase for this was "selection by consequences." Once again, variation, selection, and retention—but applied to a different stream of organization.

The fourth mechanism of inheritance is cultural evolution. When learned behaviors are copied by others (social learning), they can become transgenerational. Cultural evolution takes place in many species, including so-called lower vertebrates, such as fish, in addition to so-called higher vertebrates, such as primates. However, our species is clearly in a class by itself when it comes to cultural evolution, in large part because of our capacity to use symbols to represent objects, people, and events that aren't physically present, as well as the new forms of retention—this very book, for example—that it affords. We will look more closely at how this capability redefines behavior, including collaboration, in chapter 4.

Cultural evolution allows us to rapidly adapt to entirely new environments. It is the cultural evolution of symbolic systems that enabled our ancestors to spread over the entire planet, adapting to all climatic zones and dozens of ecological niches. Then, with the advent of agriculture, cultural evolution favored ever-larger societies, leading to the largely cooperative megasocieties of today. Note that this was not a linear progression but a tug-of-war between levels of selection at various scales. Empires rose on the strength of cultural group selection between groups and fell on the basis of disruptive cultural evolution within groups.

All of these mechanisms of evolution interact. Behavioral changes in farming practices led to the ready availability of milk in some cultures, which, over time, led to genetic changes in the capacity of adults to digest lactose. Individual learning leads to preferences for certain environments (called niche selection) that then leads to bodily changes via genetic and epigenetic evolution. It is because the flamingo could seek out the right riverbeds to dig for tiny mollusks (a learning process rewarded by the tasty treat) that its odd scoop-shaped beak and whalelike system of filters evolved. The selection pressure came from the environment, but it was learned behavior that brought the birds into contact with that environment. Much the same thing happened when beavers learned to build river dams and houses with underwater entrances. Their behavior produced a new niche—a long-standing change in the environment. This act of "niche construction" reduced the potential for predation and led to the genetic evolution of the anatomical traits necessary for harvesting timber—those great teeth! In this fashion, the slow process of genetic evolution follows wherever the faster processes of individual learning and cultural evolution lead.

This is true for many species, but it's especially true for us humans—and for a simple but profound reason. We evolved in small bands and groups. And that changed everything.

Major Evolutionary Transitions

The tug-of-war involving within- and between-group selection is not static. Sometimes mechanisms can evolve that suppress the potential for disruptive selection within groups, and thus between-group selection becomes the dominant evolutionary force. Said more simply, when the selfishness of a smaller level of selection is reined in, the benefits of cooperation at a higher level can become a relatively powerful source of evolutionary development. When this happens, occasionally something magical occurs: the group evolves to be so cooperative that it qualifies as an organism in its own right. This is called a major evolutionary transition.

The idea that individual organisms might evolve from highly cooperative *groups*, rather than from the small mutational steps of other *individuals*, was beyond Darwin's imagination. It was first proposed in the 1970s to explain how nucleated cells (called eukaryotic) evolved as a result of the symbiotic associations of bacteria, and not from the small mutations of bacterial cells (called prokaryotic). This idea was then generalized to explain other major events in the history of life, possibly even the origin of life itself in the form of groups of cooperating molecular interactions. The transition to multicellular organisms required that disruptive competition among cells within the organism (for example, cancer) be suppressed. In order for ultrasocial insect colonies to evolve, such as those of ants, bees, wasps, and termites, a similar process had to occur. The disruptive competition among individual insects within the colony had to be suppressed in order for the colony itself to be the unit of selection. In each of these cases, the selection characteristics of the higher-level units as a whole shaped the behavioral and ultimately genetic characteristics of the lower-level units.

It's tempting to equate an organism with something that is physically bounded, such as a single-cell or multicellular organism, but ultrasocial insect colonies show that physical boundedness is not a requirement. On any particular day, the members of a beehive may be dispersed over several square miles, but they still coordinate their activities through social interactions for the benefit of the whole colony, based on millions of generations of colony-level selection. They even share a group brain! For example, when a honeybee colony becomes too large, scout bees fly from the swarm in search of empty cavities. Exquisitely detailed research by Thomas Seeley and his colleagues has shown that the swarm is as discerning as a human house-hunter in making the right choice based on factors such as size, height, orientation to the sun, and size of the opening.[8]

How do the bees decide where to create a new colony? It happens through a social process that takes place on the surface of the swarm, in which each bee plays such a small role that it is more like a neuron in the hive's collective "brain" than a decision maker in its own right. We put "brain" in quotation marks, but that's truly what it is. After all, what is the brain of a multicellular organism but a group of neurons interacting socially in the right way to make good collective decisions? The vertebrate immune system is even more similar to a social insect colony than is the vertebrate brain, because its many specialized cell types communicate and coordinate their activities while dispersed throughout the body. Ultimately, it's clear that what defines an organism is not the distance between members of a group, but the level of selection at which they operate!

The concept of a major evolutionary transition is new, but the idea that human evolution qualifies as one is newer still, having emerged near the turn of the twenty-first century. Considering human evolution in this way, however, explains nearly everything that is distinctive about our species.

We are different from other species in three main ways. First, we easily cooperate with others who are genetically unrelated to us and are by far the most cooperative primate. Second, our cognition is different, especially our capacity for symbolic thought and relational learning. And, third, we culturally transmit more learned behavioral information across generations than other species. Cooperation, cognition, and culture can be called the three Cs of human distinctiveness.

And it seems likely that the three Cs evolved in that order. Our distant ancestors managed to suppress disruptive self-serving behaviors within groups, so that the group became the primary unit of selection. Physical and mental cooperation appears to be our first signature human adaptation. Our extraordinary levels of cooperation then probably provided an environment in which symbolic language, and therefore cognition, could evolve. Symbols allow more efficient cooperation, but we would not have evolved to use them unless individuals were already interacting to achieve goals. And, of course, language and cognition then allowed for culture, an inventory of symbols with shared meanings and the transmission of learned information across generations using those symbols, which requires cooperative interactions among learners and teachers.

Thus, all three Cs of human distinctiveness are, ultimately, forms of cooperation. A *single* increase in cooperativeness during human evolution can account for nearly *everything* that is distinctive about us. This is sometimes called the "cooperation came first" hypothesis.[9]

The degree of our cooperation has been increasing ever since. With the invention of the Internet, enabling a global "nervous system" of communication, we might just be on the cusp of a major shift toward unprecedented levels of cooperation. But to get there, we are going to have to learn how to honor individualism at the same time we strive for greater and greater cooperation. To reach this future, we need to understand the importance and power of small groups and, in particular, ways to bring together groups of small groups to form effective networks of interaction.

As we'll discuss in the next chapter, we think that humanity's next major evolutionary transition needs to move us beyond simplistic ideas of either top-down regulation or laissez-faire bottom-up competitive markets. It needs to move us toward forms of social organization that preserve individual self-determination but also have embedded within them evolutionarily consistent principles of design. We believe the evolutionary view of life, particularly as it has been expanded in recent years to include multiple levels and multiple streams of evolution, is an essential precursor to a deep understanding of the processes of cultural change needed to get us there.

Conclusion

This chapter covered some important ground. First, we reviewed the basic processes of evolution: variation, selection, and retention. Then we explored how selection occurs at multiple levels and in multiple streams, including genetic, behavioral, and cultural. These levels and streams interrelate in both bottom-up and top-down ways. All of the fast-paced changes swirling around us and even within us are evolutionary. A key principle to take from this chapter is that selfish behavior at one level can interfere with effective cooperation at higher levels.

We introduced the idea of major evolutionary transitions, in which selfishness at the lower level is successfully suppressed to such an extent that the group in question becomes a new kind of organism in its own right. This concept looks very different when considering human societies versus termite mounds. For one thing, our capacity to use language to form separate identities and goals means that we would never be satisfied with the kind of sameness required of termite workers. For another, humans can easily move between groups, whereas mitochondria in a cell have no chance of escape. As we will see throughout this book, these realities mean that we need to encourage cooperative behaviors within and among our groups as much as we need to work on suppressing self-interested ones.

Evolutionary theory offers a novel perspective on virtually all efforts to accomplish positive change at all scales, from that of the individual to that of the planet. One key insight with this theory is that there is something special about small groups, which were the only human social environment for most of our evolutionary history, and, like the cells of a multicellular organism, they need to remain the building blocks of modern large-scale societies. If you are already a member of a group that functions well to accomplish worthwhile goals, then you know how fulfilling it can be. Conversely, if you are socially isolated or are a member of a group that's riven by conflict, or has goals you no longer believe in, then you know how frustrating working within a group can be. In the next chapter we will introduce you to the work of Elinor Ostrom, who received the Nobel Prize in economics for her work exploring the core design principles that underpin successful cooperation within and between groups.

CHAPTER 2

Elinor Ostrom and the Commons

On December 10, 2009, Elinor Ostrom strode across the stage at the concert hall in Stockholm to receive her Nobel Prize in economic sciences from the king of Sweden. The first woman to receive the prestigious award, she was being recognized for challenging some of the most basic economic assumptions about who we are and how we might best cooperate. She worked primarily with people attempting to manage common-pool resources, such as forests, pastures, fisheries, and irrigation systems, and she framed her ideas in the language of her academic discipline of political science. David was privileged to work with her for three years, prior to her death in 2012, to frame her ideas more generally in terms of evolutionary theory and in a way that can be applied to all kinds of groups.

It would be hard to find a more down-to-earth Nobel laureate than Ostrom, who insisted that everyone should call her Lin. Without much encouragement from her parents, she put herself through UCLA and eventually earned a PhD in political science. She married Vincent Ostrom, one of her professors at UCLA, and followed him to a new position at Indiana University. There was no job there for Lin, but she made herself so useful that she was eventually asked to become the graduate advisor for the political science department. Only after her work there did they realize that they needed to offer her a tenure-track position! Once on that firm academic footing, she began to address the fundamental question that would earn her the Nobel Prize: How do groups managing publicly accessible, finite resources that are vulnerable to being depleted avoid the binary into which so many states and governments fall, of the top-down regulation of resources versus full privatization? The core design principles Lin codified from her research are central to the Prosocial process; they provide the steps groups can follow to balance individual and collective interests in a prosocial way.

The Tragedy of the Commons

In 1968, the ecologist Garrett Hardin published the article "The Tragedy of the Commons" in the prestigious journal *Science*.[1] The article has since been cited more than 35,000 times and has become a staple of first-year economics classes, in which students are taught that it reveals a fundamental truth of human nature. Part of the reason the article has been so successful, and part of the reason it has had such a huge impact upon public policy, is its deeply evocative prose. Hardin wasn't making a technical argument, he was making a moral one. And to do that he knew he had to speak to the heart not just the head. His parable goes like this:

> Picture a pasture open to all. It is to be expected that each herdsman will try to keep as many cattle as possible on the commons. Such an arrangement may work reasonably satisfactorily for centuries because tribal wars, poaching, and disease keep the numbers of both man and beast well below the carrying capacity of the land. Finally, however, comes the day of reckoning, that is, the day when the long-desired goal of social stability becomes a reality. At this point, the inherent logic of the commons remorselessly generates tragedy...
>
> The rational herdsman concludes that the only sensible course for him to pursue is to add another animal to his herd. And another... But this is the conclusion reached by each and every rational herdsman sharing a commons. Therein is the tragedy. Each man is locked into a system that compels him to increase his herd without limit—in a world that is limited. Ruin is the destination toward which all men rush, each pursuing his own best interest in a society that believes in the freedom of the commons. Freedom in a commons brings ruin to all.[2]

Hardin was using "commons" to refer to "common-pool resources," which are resources such as forests, pastures, fisheries, and aquifers that are publicly accessible and finite and therefore vulnerable to depletion. Conventional economic wisdom, based on Hardin's logic, has held that the only way to prevent the tragedy of the commons is to either privatize the resource and use markets when possible, or for states to impose top-down regulation.[3] Hardin himself favored the latter.

Ostrom recognized that top-down control can work up to a point, but governments lack key capabilities. It is often hard for top-down regulators to know the exact capacity of a resource, to monitor and appropriately respond to infractions, and to provide the most engaging conditions for individual action. With top-down regulation, Ostrom's research showed that people tend to "defect" by breaking or bending the rules. Furthermore, such an arrangement is inefficient because the "controller" is outside the system being affected yet is consuming resources without creating benefits.

As for privatization and the market, private property rights only work when the owners actually care for the resource. If instead their priority is to exploit the resource as rapidly as possible before moving on, then private property does little to help the long-term management of common-pool resources. Furthermore, by their nature many common-pool resources—for example, the atmosphere—cannot be privatized.

Ostrom argued that we should not be drawn to the simplicity of either of these two models, but instead take on the difficult, context-bound work of designing approaches in which those most affected by the issue (that is, the herders, farmers, or fishers) make binding agreements to commit themselves to cooperation. Her PhD thesis documented a success story involving unregulated groundwater use in Southern California, where the water table was rapidly declining as a result. If it sank too low, there would be an incursion of seawater, which would have been cataclysmic for everyone. The stakeholders

Ostrom studied—various municipalities, farmers, and the oil industry—found a way to manage the resource on their own, cobbling together an agreement to preserve their precious common resource.

She then studied the problem more comprehensively, including by reviewing common-pool resource groups in traditional and industrial societies around the world. From the review she concluded that groups *are* able to manage and sustainably share their common-pool resources, without privatization or top-down regulation—without needing others to enforce what they did—but only if they enact certain design principles to forestall the tragedy of overuse and other failures of cooperation. This insight was so new against the background of standard economic wisdom that it earned her the profession's highest honor, the Nobel Prize.

Here are the core design principles that Ostrom derived for common-pool resource groups that enabled them to avoid the tragedy of the commons:

1. Groups that functioned well had *clearly defined boundaries*, which means that they knew they were a group, what the group was about, and who was a member.

2. They also had rules adapted to local conditions, which ensured that group members experienced *proportional equivalence between benefits and costs*.

3. There were *collective choice arrangements* so that everybody affected by a decision had an opportunity to contribute to making that decision.

4. There was *monitoring*, which means that lapses in agreed-upon behaviors could be detected; and this monitoring was preferably conducted by group members rather than external authorities.

5. There were *graduated sanctions*, which means that lapses in agreed-upon behaviors were corrected, gently at first but with a capacity to escalate if necessary.

6. There were *conflict resolution mechanisms* that quickly resolved differences in a way recognized as fair by all members.

7. There was at least a *minimal recognition of rights to organize*, which gave group members the authority to manage their own affairs without external interference, the opposite of top-down regulation.

8. Finally, in the case of larger common-pool resources, there was *polycentric governance* (literally "many-centered" governance), which refers to groups of groups in multiple layers of nested enterprises, with each subgroup relating to other groups using principles 1 through 7. In other words, successful common-pool resource groups had prosocial relations with other groups.

Take a moment to reflect upon these core design principles (CDPs). There is nothing esoteric about them. They seem like common sense. And any group might benefit from them, not just groups managing common-pool resources. But if these core design principles are so effective, easy to understand, and universal, then why don't all groups adopt them?

Why Aren't the Core Design Principles Used Everywhere?

The groups Ostrom studied employed the CDPs differently, and some of the groups adopted them, whereas others did not. It's because of this variation that she was able to derive the CDPs in the first place, by looking at the characteristics of groups that worked. Perhaps you, too, can attest to the frequent absence of one or more of these design principles in groups to which you belong? Certain groups, such as neighborhoods, often don't have a strong sense of group identity (a violation of clearly defined boundaries). Others frequently consist of a few beleaguered volunteers who do most of the work (a violation of benefits being proportional to efforts). Organizational efforts to change frequently founder due to a lack of input or engagement from people affected by decisions (a violation of collective choice arrangements), or members of a work team feel ignored or micromanaged (violations of monitoring and graduated sanctions). Discipline in schools is frequently neither fast nor based on a procedure that the students perceive as fair (a violation of conflict resolution mechanisms), and so on.

As we noted, you may see these design principles as common sense, so why are they not more purely instinctive to all groups?

We could ask the same question of other basic biological adaptations: How did the cultural practice of bottle-feeding infants become so widely established, for example, when lactation has been the signature mammalian adaptation for nearly 200 million years? Part of the answer is that for virtually all of that time, female mammals had no alternative to breastfeeding and therefore no reason to evolve a preference for it compared to an alternative. Similarly, throughout our evolutionary history, our ancestors had no alternative to living in small social groups and therefore did not necessarily evolve the instincts for creating them when alternatives became available. It is only now, faced with the enormous challenges of organizing in nationwide or even worldwide groups, that we are challenged to become more conscious of the core principles of effective cooperation.

David was thinking about this evolutionary aspect of cooperation when he met Ostrom at a workshop in 2009, just a few months before she was awarded the Nobel Prize. He realized that her CDPs approach fitted multilevel selection theory like a glove. Imagine a group that strongly implements all eight of the CDPs. Now imagine a member of that group trying to benefit himself or herself at the expense of others or the group as a whole. With the CDPs in place, that would be very difficult. The core design principle of inclusive decision making means that everybody's interests will be represented, not just a select few, and the principles of monitoring agreed behaviors and graduated sanctions mean that people would find it harder to get away with transgressing agreements, and so on. The best way to succeed in such a group is through teamwork. Conversely, imagine a group that lacks all of the CDPs. Such a group would be easy pickings for a member who cared more about his or her selfish interests than the interests of the group. In groups with the CDPs, benefits naturally flow from cooperation, whereas in groups without them self-interest can easily dominate, undermining the effectiveness of the group overall.

David wondered whether implementing the CDPs in a group might be the key to getting the group to function more like a single entity, a bit like a miniature major evolutionary transition. Perhaps groups enacting these principles could suppress within-group competition and support cooperation to such an extent that something new, a truly high-performing group, might emerge—metaphorically a new organism operating in its own ecology. He posited that if we took cultural evolution seriously, implementing the first seven CDPs effectively in a group could literally be a cultural major evolutionary transition for that group, and applying the eighth principle (polycentric governance), whereby the first seven principles are reapplied at the next level up, would allow the creation of larger ecosystems of groups. This, he thought, would be niche construction at a cultural scale, which leads us back to the purpose of this book.

Ostrom didn't win her Nobel Prize for a kind of how-to recipe for improving the internal operations of particular groups. She won it for challenging the dominant economic paradigm, and for starting to articulate the beginnings of an alternative, more hopeful vision for humanity. While most of this book will be about the practicalities of how the Prosocial process can help your specific group cooperate more effectively, both internally and with other groups, we would not be doing Ostrom's work justice if we turned it into a simple how-to guide for team development. Instead, what we want to do in the remainder of this chapter is paint a picture of how the local actions you take in response to reading this book might actually be seen as part of a larger global movement toward a more harmonious, equitable, and sustainable society.

The Rise of *Homo Economicus*

The power of the tragedy of the commons parable is that it paints a plausible and simple picture of human nature. Each herder is solely interested in maximizing the returns they get from the pasture. They have zero interest in other outcomes, such as the inherent value of the land itself, the well-being of other herders, or the long-term sustainability of their practices. In addition, each herder is an entirely independent unit. Indeed, in Hardin's version, not only is each herder blithely unaware of the ways they are connected to other herders, there isn't even any room for discussion or agreement.

The tragedy of the commons has been taught as absolute truth to generations of students, many of whom now run the political, economic, and social institutions that govern our lives. Indeed, this way of thinking came to define a whole new way of thinking about the nature of human beings, known as *Homo economicus*, or economic man—a being solely focused on satisfying their own self-interest at the lowest possible cost irrespective of the effects of their actions upon others.

In his wonderful book *Sapiens*, Yuval Noah Harari makes the point that our capacity to cooperate flexibly and in large numbers is based upon our capacity to construct shared fictions that organize our behavior. Religions, organizations, laws, human rights, nationality, and even money all exist only in our collective imagining. Corporations are legal fictions created by lawyers. You could replace every chair, document, or person at the

Apple corporation, and Apple would still exist. But it could be undone in a moment if some lawyers and a judge revoked its legal basis for existence. Similarly, human rights don't exist inside a human; they are culturally constructed to improve human relations. With this logic in mind we can easily see how states and nations exist only as long as people agree on their borders, which animals cross with absolutely no awareness of immigration laws. As to money, its value is entirely constructed. For example, a child might naturally prefer a nickel over a dime simply because it's larger. The child has to be taught that a dime is "worth" more than a nickel.

But the most important fiction of all is the one we tell ourselves about who we are—our identity. This is where the real power of the concept of *Homo economicus* resides. The psychologist Robert Frank compared economics students with anthropology students and found that after just two terms of training in the economic view of life, economics students were more likely to believe that others act selfishly and, most importantly, they were less likely to help others in situations where they had the opportunity to do so.[4] If we see others as exclusively self-interested, we naturally act to protect ourselves and act in our own interests. The more we argue that we are self-interested, the more self-interested we become. *Homo economicus* is, in other words, a self-fulfilling prophecy.

A recent study of one thousand people from Britain by Common Cause revealed some extraordinary implications of the way we think about ourselves.[5] While 74 percent of participants valued compassionate values over selfish values, 77 percent underestimated the degree to which others held compassionate values and overestimated the degree to which others held selfish values. In a subsequent part of the survey, Common Cause asked about the values espoused by our social institutions, such as government, the media, the educational system, and business. On average, people thought that all these social institutions espoused more selfish values than they, themselves, held. And these beliefs about others had real power, just as they did in Robert Frank's studies. The people who inaccurately believed that others were more selfish than they were didn't get as involved in meetings, voting, or volunteering. They didn't take as much responsibility for the functioning of their communities, didn't feel like they fit into society, and were more likely to be socially alienated. If I want to be prosocial, but I believe others are selfish, I am less likely to take the risk of contributing to the welfare of others and much more likely to focus on getting my own needs met.

After many decades of working with people in groups, we have seen repeatedly how assumptions about human nature shape the way that people relate to others. A manager who thinks that his workers are basically selfish and will work as little as possible will get very different outcomes than a manager who recognizes that, with the right norms and practices in place, it is possible to create an environment in which people want to give their all on behalf of the team.

Our view of human nature not only changes what we do personally, it changes the forms of governance we design and implement. If we are all separate and selfish, then we need to either force people into suppressing their self-interested behavior through coercive rules and regulations, or we need to seek to exploit that selfishness by giving market-based competitive forces free rein. We are right now in the midst of seeing worldwide what happens when just these two options prevail, which we'll explore in the next section.

Polycentric Governance: Beyond Regulation and Markets

Throughout her career, Lin Ostrom collaborated closely with her husband Vincent. Arguably their single greatest contribution was to reempower small groups. They recognized that by applying the first seven core design principles at ever higher levels of organization, small groups could be the bedrock of much larger polycentric (that is, many centered) systems of governance.

Polycentric governance respects the localized needs, knowledge, and autonomy of small groups, *as well as* the need for modern societies to organize at a regional, national, or even global scale. It takes into consideration that life consists of many spheres of activity, that each sphere has an optimal scale, and that good governance requires identifying the optimal scale for each sphere and appropriate coordination among the spheres. We can look to Ostrom's own work for an example of optimal scale. In metropolitan police departments, it is optimal for the cops walking and driving the streets to be quite local, so that the people in the neighborhoods they're patrolling can come to know them. But forensic labs should be regional, to avoid redundancy in the purchase of expensive equipment.[6]

When one thinks about it, the concept of polycentric governance has the same obviousness as do the CDPs for single groups. Yet, that doesn't mean that it is widely employed! There are two major forms of large-scale governance that violate the principles of polycentric governance and are also diametrically opposed to each other: laissez-faire (consisting of privatized resources controlled within markets) and centralized planning (consisting of top-down regulation, say by a government or a group of experts).

The concept of laissez-faire (French for "let it be") dates back to the beginning of the economics profession and assumes that the unregulated pursuit of lower-level self-interest (for example, individuals or corporations) robustly benefits the higher-level common good. It is expressed in Adam Smith's metaphor of the invisible hand and Bernard Mandeville's *The Fable of the Bees*. French economist Léon Walras generated a mathematical formulation to explain it in the nineteenth century, which developed into rational choice theory in the twentieth century. Ayn Rand gave it artistic and moralistic expression in her novels and philosophical works, such as "The Virtue of Selfishness." It provides the justification for the neoliberal economic policies that became dominant during the Reagan and Thatcher eras in the United States and United Kingdom and is the dominant economic model used today globally.

Conversely, centralized planning assumes that a group of experts can figure out and implement the best forms of governance. The practices and beliefs of the Keynesian branch of the economics profession, of nations such as communist Russia during the mid-twentieth century, and of many (but not all) business corporations generally reflect this concept. It is ironic that many business leaders who champion laissez-faire for the economy insist upon "command and control" for the governance of their own corporations!

Both approaches have strengths that an improved form of governance should retain but also grave weaknesses that need to be corrected. The main weakness of laissez-faire is that the unregulated pursuit of lower-level self-interest does *not* robustly benefit the common good. If this is not obvious today based upon the state of our world, then MLS theory makes it obvious that adaptation at a given level of social organization requires a process of selection at that level and tends to be undermined by selection at lower levels. Even the most die-hard libertarian admits that some regulations need to be imposed upon laissez-faire capitalism. Market failures have led to extreme inequality and potentially catastrophic environmental degradation. Markets privilege the individual's desire over the collective's long-term interest, and giving markets free rein marginalizes any idea of working together toward common goals. *Homo economicus* views people as rational actors focused on satisfying their own self-interest at the lowest possible cost, irrespective of how this activity affects others. Free-market adherents have not only accepted this view but celebrated it as the path toward human thriving—and society's sense that there is a collective, common good that we must preserve through governance has been, as a result, significantly impoverished.

Centralized planning, on the other hand, has at least three major problems. First, human social systems are too complex to be understood by a team of experts. This is even true for a moderately sized business corporation, not to mention a nation or a global economy. Outside regulators often have limited information compared to those who actually manage a common-pool resource or work in a particular group—and, as Ostrom's third core design principle (collective choice arrangements) shows, whenever the people creating the rules are separated from those affected by them, the likelihood that decisions will be relevant, timely, and effective is much diminished. Second, those higher up in top-down hierarchies are often tempted to act more from self-interest than the collective interest of the group. Centralized planning can exacerbate rather than suppress selfish individualism. And, third, centralized planning has damaging effects upon those who are regulated. Hierarchies of power all ultimately rest upon coercive powers. Just as nations have armies to impose their will on other nations, and police forces and legal and penal systems to impose their will on their own citizens, so too do managers in strongly hierarchical organizations have power to coerce their staff. But one of the most fundamental needs of human beings is to be able to endorse one's own actions as volitional. And literally thousands of empirical studies have shown that vitality and creativity are dampened when people's behavior is controlled by the fear of retribution, by compliance, and by the need to seek the approval of authorities.[7]

Both of these views have taken our attention away from a much older form of governance, that of the commons. Ostrom's work has brought the attention of many to the commons, and with the rise of the Internet we are seeing new forms of cooperation springing up around the world that paint a more hopeful image of humanity. It is to this idea that we now turn.

The Modern Commons: Anything but a Tragedy

Most of us think of a commons as just a resource, such as a field or a park. This was certainly the way that Garrett Hardin used the term. But commons scholars have argued that Hardin wasn't really talking about a commons at all, but "an open access regime or free-for-all in which everything is free for the taking."[8] Ostrom rebutted Hardin by documenting situations where communities created social practices that successfully managed access and avoided overexploitation. Common-pool resources such as lands, bodies of water, the atmosphere, space, and even the Internet only become the commons when they are actively governed by communities that care for their sustainability. The commons, according to David Bollier, "is a resource + a community + a set of social protocols. The three are an integrated, interdependent whole."[9] The community has an interest in managing the resource, and the social protocols are intended to help manage the resource for the benefit of all members of the community. As you will see, this emphasis upon human valuing and social agreements is at the heart of the Prosocial process, at the scale of multigroup society and single groups.

The word "commons" derives from the Norman word *commun*. To *commune* is to participate, to share and enjoy fellowship. And in turn the word "commun" draws upon the word *munus*, which means "a gift, service, or duty" that in turn contributes to the word *munificent*, meaning "bountiful, liberal, generous," and the Latin *munificare*, meaning "to enrich." Thus the "commons" literally refers to sharing gifts to enrich all. Sharing of gifts implies both the satisfaction of individual needs and a duty to the collective. David Bollier argues that it is more helpful to treat the commons as a verb rather than a noun: "commoning" rather than "the commons."[10] *Commoning* is the care the community feels for the resource and the community itself; it is the making and implementation of rules of access and use; and it is the social norms and customs that evolve to ease and resolve conflict, encourage cooperation, and punish free riders and shirkers. Indeed, what Bollier calls "social protocols" is what Ostrom sought to codify in her CDPs. And the Prosocial process is our attempt to embody the ethic of commoning in a set of actionable steps.

There are millions, possibly billions, of commons efforts around the world. With the rise of the Internet, examples of commoning span a wide range of forms, from physical commons, such as community gardens and tool-sharing systems, to vast virtual collaborations; from small local groups, such as toy-lending libraries, to global platforms, such as microlending facilities or house sharing. We are seeing extraordinary feats of cooperation that would have seemed impossible just a decade or two ago. Let's review a number of examples, so you can sense the power that could come from managing evolutionary processes at the level of small groups.

Wikipedia is the world's largest global database of knowledge and is built on the voluntary shared effort of people who care about information. Uber, Lyft, and numerous other platforms enable shared rides. Airbnb and Couchsurfing have made it possible for people to share their most private of spaces, their homes. Open-access journals make knowledge available to all, to even those in developing countries, unlike closed pay-to-view journals. The Conversation is a journalistic commons where academics from Australia and around the world publish works on topical issues, with republishing encouraged under a creative commons attribution/no derivatives license.

Entire cities have moved to rebuild the commons. The Bologna Regulation for the Care and Regeneration of Urban Commons has reenergized the city (based on Ostrom's ideas) by empowering its citizens to design spaces and programs in their city.[11] The city government invites citizens to partner with it to conduct joint "public-commons" management. Dozens of projects have been conducted to transform parks and abandoned buildings and schools, and more than 140 other cities have enacted similar projects. This approach has shifted the emphasis from citizens as consumers of government services to citizens as policy makers that the city enables and supports.[12] The city of Ghent in Belgium has created more than 500 commons, including food commons, community land trusts, renewable energy cooperatives, and neighborhood-managed public buildings. These cities are classic examples of small groups organized around projects that in turn transform a larger group—an entire city.

Each of these local initiatives strengthens the social capital of its respective city. For example, two young women recently decided to use the Internet to organize a system of "welcome dinners" for new arrivals to Australia. The idea is simple. People who live in a town or region register their interest in hosting or attending welcome dinners, and refugees, foreign students, and other new arrivals register their interest in attending a dinner. Everyone brings a dish to share, and there are usually some brief informal activities to break the ice so people can get to know one another. In the simplest of ways, the Internet enables real human connection in the form of a delicious international meal. Everyone leaves with smiles on their faces, enriched by both the cultural exchange and the knowledge that they have engaged in the oldest of human pastimes: meeting, laughing, and talking while sharing a meal. The welcome dinner project nicely illustrates the ways in which the Internet, in particular, can smash through issues of scale. It is deeply local but also global; it makes use of economies of scale to deal with shared issues, such as ensuring that the dinners are safe and secure; and it brings people together across national and cultural boundaries. Other initiatives that build sustainable communities include the slow-food movement and the transition towns movement.

Mutual aid networks are location-specific, nested cooperative structures that "create means for everyone to discover and succeed in work they want to do, with the support of their community."[13] Local mutual aid networks encourage people to create initiatives to achieve common goals, such as improving community and sustaining the environment. The groups then use tools, such as shared facilities (for example, libraries and laundries) and time banking and other exchange systems, to provide cooperative savings and lending resources and to support the cooperative ownership of resources.[14]

Other groups are recreating the commons to develop and share knowledge. Peer-to-peer universities make use of "learning circles" that enable people to learn a wide range of courses in small groups with their neighbors. Most of these groups meet in libraries and community centers in more than 111 participating cities.[15] In Australia, the University of the Third Age (U3A) brings together older, knowledgeable volunteers who create and edit courses to educate older people or younger disabled people around the world who cannot make it to traditional classes. U3A is once again organized as a series of smaller, local communities that collaborate and share resources nationally.

Other initiatives are designed to support locally based production. For example, Wikihouse was developed to allow anyone to design, share, fabricate, and assemble their own house. The platform allows people to share contextually appropriate blueprints for houses made from plywood that then fit together jigsaw style. The idea is to share what is "light" (designs, help manuals, and support) globally while what is "heavy" (the procurement of wood and the actual assembly of the house) is handled locally. The community is dedicated to continuous improvement, and, critically, it engages small local groups by creating chapters that sign agreements to preserve equity, sustainability, sharing, and self-determination.[16]

Community gardens are forming all around the world as local people coordinate behavior to meet their food needs and form a community. The Potato Park in the Cusco Valley of Peru is dedicated to preserving the local environment and nurturing the diversity of local crops. Quechua natives are legally entitled to steward the biodiversity of the region, and they sustain 2,300 varieties of the world's 4,000 different sorts of potatoes. This region thrives because localized agricultural and social practices, such as bartering, that reflect the indigenous spiritual traditions and cultural values are preserved in small groups cooperating in an entire region.[17]

Such commons can be deeply integrated with markets. For example, Linux operates on more servers than any other operating system. It was developed by programmers who were motivated to contribute and share their efforts and were organized in cooperative groups. Sensorica is a collaborative network focused on designing and building sensors and sense-making systems. It offers support to individuals and organizations that collaborate, and revenue from sales is distributed back to the network proportional to member contributions. Agreements exist to prevent individual interests from dominating at the expense of the collective, and no agent can have dominant control over the shared resources.[18]

Enspiral is another inspirational initiative in this respect. It consists of a network of member-designed and member-run businesses that operate as separate small groups (businesses) but exist within an ecosystem of shared resources. The Enspiral Handbook is itself a kind of public commons that articulates the agreements, norms, and practices that sustain this extraordinary venture.[19] Enspiral is a classic example of bringing together individual needs and passions with the benefits of collaborating in larger groups.

These are just a few of the countless examples springing up globally. What can we learn from these efforts? First, all of these examples successfully balance self-interest with the broader needs of the collective by simultaneously empowering individuals within the context of a collective vision.

Second, all the examples emphasize ongoing relationships rather than single transactions, which is characteristic of many markets. One of the most robust findings from social dilemma research is that people who anticipate an ongoing relationship are much more likely to be prosocial than those who do not. This is why ongoing small groups are important—we need to attend to our stable relationships, not just see ourselves as isolated units in a massive undifferentiated pool of other isolated units.

Third, all of these examples operate as complex adaptive systems, preserving wherever possible the individuality of constituent units, the rich tapestry of cultural practices, and local environmental conditions, histories, and stories. While global, these efforts do not seek to reduce relationships to a single market logic, in which everything is a commodity for sale. Instead, commoning not only recognizes the uniqueness of each situation, it welcomes the uniqueness as an essential aspect of building communities that care.

Fourth, many of these modern examples of the commons manage to optimally combine the local scale with the global, which is exciting. As you will see in chapter 14, when we further elaborate on the eighth design principle, this is a key characteristic of polycentric governance.

Commoning is always local and contextualized. It always exists in a context, and as such the pragmatics of how best to manage a resource in a sustainable way inherently drive it, rather than preexisting beliefs, theories, or ideologies about how management should occur in general. Sensitivity and flexibility to context is a key aspect of human thriving. As Bollier puts it, "The commons works because people come to know and experience the management of a resource in its unique aspects. They come to depend on each other and love this forest or that lake or that patch of farmland."[20]

Whereas selfishness, separation, transactionalism, and a kind of one-dimensional valuing is entrenched in *Homo economicus*, commoning invites connection and privileges values such as inclusiveness of participation, equitable access and use, bottom-up empowerment, accountability for individual behavior, and long-term sustainability for the mutual benefit of all. Commons-based organizations rely upon a much greater range of values and purposes than just maximizing market value. The community's needs, even the needs of ecology, the degree of self-determination, social equity, the quality of relationships, the richness of culture, and connections to place can all take precedence over the simple idea of financial accumulation.

In Conclusion: Prosocial as Applied Cultural Evolution

If we step back and look at *Homo economicus* through the lens of evolutionary theory, we quickly see that it is a narrative that has shifted the balance of what we have selected for as a species, from cooperative groups to individualistic, selfish behavior. To shift the balance back, creating miniature evolutionary transitions toward highly cooperative groups, we need to construct new niches for groups, new contexts in which they can operate that support cooperation and diminish the likelihood of selfishness. We need to construct conditions in which altruism can thrive. Changing the narrative is part of that because it changes how we expect others to behave. If we expect others to behave selfishly, we are more likely to behave so ourselves. If we expect others to behave benevolently, we are more likely to feel safe behaving so ourselves.

Ostrom's principles can help us make this shift, which is why we wanted to focus on them while creating a new framework for effective cooperation within and between groups. Monitoring, sanctions, and effective conflict resolution make it safe for us to call out and seek to resolve uncooperative behaviors. Inclusive decision making and proportionality of benefits to contributions motivates effort and discourages free-riding behaviors. Shared purpose and identity create the very ground for collective action. And being able to do all this without excessive interference from outside protects the group from inappropriate constraints that are not locally adapted.

For coordinating on a wide scale, Ostrom's work shows us an alternative to either laissez-faire or centralized planning. This third form of large-scale governance—commoning—is more evolutionary in nature and is based on a collection of groups cooperating with other groups at scales that are optimal for multiple spheres of activity. Commoning has arisen by cultural evolution many times, but typically in isolation and without a general theoretical framework to identify its key principles, which are necessary to deliberately and successfully instill this concept.

The power of Ostrom's principles is not that they are new. As we have noted, they are so regularly discussed as to seem almost obvious. Their power is in their generality, which allows them to be applied at any level of cultural evolution. This means that the Prosocial process is just as useful for helping to improve the internal functioning of a team as it is for improving the ways that different organizations cooperate within an innovation ecosystem and even, potentially, the ways that nations cooperate on the global scale.

CHAPTER 3

Core Design Principles, Version 2.0

Stephen Covey once said, "There are three constants in life…change, choice and principles."[1] We want the Prosocial process to be useful for as many different kinds of groups as possible—a couple or family, a classroom, an aid group providing community development services, a corporation, or a soccer team. We want the principles to guide the creation of shared social practices that help groups dynamically balance individual and collective interests. We want groups to function so that individuals can be authentic and free but exist within strong, caring groups that leverage their effectiveness and meet their social needs.

In this chapter we provide a generalized version of Ostrom's principles that can be applied to any group, or group of groups, that wants to be more cooperative. Brilliant as Ostrom's work was, it was primarily descriptive and was focused on groups managing common-pool resources. Our challenge is to take the essence and intent of her work and broaden its scope to articulate a practical way of applying the principles to all kinds of groups to *improve* them. We did not undertake this work lightly, and we remain enormously glad that David had the chance to work directly with Ostrom, so we had a strong sense of how she, too, might have undertaken this challenge.

Good principles for action, like good scientific theories, should have scope, precision, and depth. That is, they should ideally be generally useful for many circumstances but not so general as to be unhelpfully vague, and they should not be contradicted by knowledge of other levels of organization. Our articulation of Ostrom's principles is intended to be specific and precise enough to be useful, but not so specific as to only apply in a small range of situations. And we have worked hard to preserve Ostrom's essential insight that a small group of principles might be applicable at many levels, specifically with individuals within groups and with groups within larger groups.

Table 3.1 displays the Prosocial version of each of Ostrom's principles, along with the functions each principle serves within and between groups. We have preserved core design principles 3, 4, 6, 7, and 8 in more or less the original form articulated by Ostrom. In the sections that follow, we briefly explore how we've adapted the remaining principles of Ostrom's work.

Table 3.1

Ostrom's Principle	Prosocial Version	Function
1. Clearly defined boundaries	Shared identity and purpose	Defines group
2. Proportional equivalence between benefits and costs	Equitable distribution of contributions and benefits	Ensures effectiveness by balancing individual and collective interests
3. Collective choice arrangements	Fair and inclusive decision making	
4. Monitoring	Monitoring agreed behaviors	
5. Graduated sanctions	Graduated responding to helpful and unhelpful behavior	
6. Conflict resolution mechanisms	Fast and fair conflict resolution	
7. Minimal recognition of rights to organize	Authority to self-govern (according to principles 1–6)	Ensures effectiveness while supporting engagement
8. Polycentric governance	Collaborative relations with other groups (using principles 1–7)	Scale to entire systems

Core design principle 1 functions to define the group. Without shared identity and purpose, there is no possibility of being a cooperative group. But this principle is only necessary and not sufficient for cooperation to occur. Principles 2 through 6 balance self-interest with collective interest, seeking to suppress selfish individualism while encouraging creativity, individual engagement, and cooperation. Principle 7 functions to preserve the capacity of the group to manage itself. As Ostrom realized, when groups are excessively constrained in the ways that they manage principles 1 through 6, they are less likely to appropriately adapt the principles to their contexts. Psychological research further indicates that groups are less likely to engage the principles when the group's self-determination is impaired. Finally, principle 8 is not really about cooperation within a group at all. Instead, it points to the fact that the first seven principles are replicable at higher levels of organization. So groups get along better with other groups when there is shared purpose, equity, inclusivity, transparency, responsiveness, effective conflict management, and appropriate levels of self-management *between* groups as well as within them.

We see the principles as pointers that guide the attention of groups toward what works. In this sense, principles can variously look like

- valued qualities of behavior, such as equity and inclusiveness;

- broad patterns of behavior, such as clarifying who is in the group or out of it; or

- desired outcomes, such as group members feeling valued by others in the group, and a shared sense of accountability.

In the remainder of this chapter we provide a brief overview of all of the principles, enough for you to have a sense of the ways that the principles work as a whole and what behaviors we're trying to shift before we talk about the science of behavior change in the following chapter. We will dive much more deeply into why the principles are important, and how to apply them, in chapters 7 through 14.

Core Design Principle 1: Shared Identity and Purpose

A group functions best when its members clearly understand its purpose and perceive that purpose as worthwhile. A group also functions best when it offers a strong shared identity that helps group members understand who is in or out of the group, increases pride and pleasure in belonging, and guides and coordinates behavior through shared norms and values.

For Ostrom's original study groups, the shared purpose was always the sustainable management of a resource, and so the key point of principle 1 was making sure that groups knew who was entitled to claim the resource and to create agreements to manage that resource. That is why her original core design principle 1 focused on the issue of membership—who was in and who was out of the group. But, as we described in chapter 2, a more general view of the commons defines it in terms of shared purpose, not just a shared resource. Furthermore, in many groups just being a member is not enough. I can be a member of a group but still undermine it if I don't really believe in its value or care about its other members. So, the general form of membership is shared identity. Thus, when David worked with Ostrom to generalize the principles, they decided to reframe principle 1 as "shared identity and purpose."

Key values: Purpose, effectiveness, belonging, caring.

Key behaviors: Continuous informal and formal consideration and use of purpose in group activities. Clarify who is in the group (and who is not). Establish clear criteria for membership.

Key outcomes: Members understand and believe in the value of the purpose of the group. Members have a shared sense of identity, both as a sense of belonging and of caring for others in the group.

Key assessment question: To what extent do group members feel a sense of belonging and shared purpose with the group?

Key planning questions: What do we most care about? How can we create a sense of caring, belonging, and safety in the group?

Example methods: At its core, implementing principle 1 involves continual reflection on purpose and the use of group purpose as a guide for daily action. Members can facilitate reflection by working with strengths-based questions, such as "What are we doing at our best?" or using tools such as the collective matrix (which we describe in detail in chapter 5). Outputs include mission, vision, and values statements, though these are less important than continual reflection on the direction in which the group is going. We describe principle 1 in detail in chapter 7.

Core Design Principle 2: Equitable Distribution of Contributions and Benefits

For groups to function well, the effort and other forms of contribution required of its members, and the benefits of that effort, need to be distributed fairly. Most people have a strong sense of equity, and it is violated when someone receives benefits disproportionate to their contributions, so perceived fairness—a fair balance between effort or workload and reward—is essential for good group performance. Sometimes, in groups that fall short in implementing this principle, perceived unfairness is "undiscussable"; sometimes it's discussed endlessly but in ways that don't lead to positive change.

Ostrom's second design principle originally had two aspects to it. First, she described how local agreements needed to be appropriate to context, and, second, she described how they needed to be equitable. Both remain important in a generalized version of this design principle. For example, some members of the Prosocial development team have professorial salaries, while others are self-employed. Receiving monetary compensation for effort is much more important for the latter than the former. In general, a good way to implement this design principle is to have a discussion with each member of the group as to what they most want to get from the group and what they have most to give. If the arrangement seems fair to everyone at the end of the discussion, then both aspects of this design principle will be satisfied.

Key values: Equity.

Key behaviors: Pay attention to individual costs and benefits and balance them appropriately.

Key outcomes: Group members feel valued and rewarded for their contributions.

Key assessment question: To what extent are the demands and benefits of participating in this group distributed equitably among its members?

Key planning question: How will we ensure fairness in this group?

Example methods: There are a variety of tools one can use to encourage the fair distribution of contributions and benefits. Begin by ensuring role clarity and transparency. From there, create channels for the open discussion of fairness and grievances; include training in communication skills to encourage assertiveness (so those who perceive unfairness can come forward) and the ability to listen to multiple perspectives (so those in management or those who may be benefiting from unfair structures can hear the concerns and respond). We elaborate on this list and the importance of this principle in chapter 8.

Core Design Principle 3: Fair and Inclusive Decision Making

If you want good decisions and motivated people, group members need to be involved in making the decisions that affect them, particularly with agreements about how the group runs. Robust evidence in psychology shows that the motivation, commitment, and long-term well-being of people comes from giving them control over their own actions and not from manipulating and coercing them to act in ways they do not willingly endorse.[2]

Key values: Individual self-determination (empowerment), making good decisions.

Key behaviors: Collective choice processes—that is, including group members in all decisions that affect them (for example, consent-based decision making).

Key outcomes: Members know their interests and perspectives have been considered fairly and efficiently.

Key assessment question: To what extent do group members feel involved in making the decisions that affect them?

Key planning question: How will we decide in a way that involves those who need and want to be involved?

Example methods: Principle 3 can take the form of formal processes of consent-based decision making, as we describe in more detail in chapter 9. But in some circumstances, such as those involving very large groups or corporate settings, consultation with a designated leader or representative, fair informal systems of voting, or even the opportunity to make objections (veto powers) can be enough and can even be more efficient.

Core Design Principle 4: Monitoring Agreed Behaviors

For groups to function well, it's essential to have some way to monitor agreed-upon behaviors. In hierarchical organizations, this type of monitoring tends to be seen as the job of a manager or other authority, but such top-down monitoring is often coercive and serves the interests of the manager, not those of the person being monitored. It's also not very transparent, and self-serving behaviors increase when there's a lack of transparency. Ostrom's work suggests that monitoring is often better performed by peers as part of the normal interaction of group members.

Key values: Transparency.

Key behaviors: Processes to enhance the visibility of behaviors of group members to each other.

Key outcomes: Misbehaviors are quickly detected at low cost to the group using processes such as pairing people with peers to whom they are accountable.

Key assessment question: To what extent do group members know what others in the group are doing?

Key planning questions: How can we be aware of what each other is doing? How can our behaviors be transparent?

Example methods: Meetings, reporting, role swapping, and other activities that help group members notice and assess what others are doing.

Core Design Principle 5: Graduated Responding to Helpful and Unhelpful Behavior

No one is perfect when it comes to fulfilling the obligations of a group. Even the most capable and well-meaning members can fail, especially given competing demands upon time and attention. Transgressions do not imply malicious intent. Effective groups have in place responses to transgressions, including open conversation to find out what happened, sanctions, or even, ultimately, exclusion from the group.

While Ostrom focused on graduated sanctions for misbehaviors—which reflects her emphasis on suppressing antisocial behaviors, a necessary component for the successful management of common-pool resources—the groups she studied tended to have stable membership, with members living in a particular area associated with the commons. That's not always the case with groups. When we generalized this principle, it became apparent that most groups require responding not only to discourage unhelpful behavior but also to encourage helpful behavior. Without positive reinforcement, there is little incentive to stay with a group, and people either disengage or leave. We'll describe this idea in more depth in chapter 11, which explores principle 5; for now, just know that our formulation of this principle is not just about sanctioning unwanted behavior, but also about figuring out ways to promote desired behavior.

Key values: Learning, compassion, competence.

Key behaviors: Responding at an appropriate level of intensity to either encourage or discourage behaviors that are contributing to or detracting from cooperation.

Key outcomes: Group members feeling valued and appreciated, feeling empowered to respond to transgressions, and having shared valuing and a commitment to high performance in the context of responding compassionately to mistakes.

Key assessment questions: If someone behaves in a way that is unhelpful or disruptive in this group, to what extent do people respond appropriately to discourage that behavior? If someone behaves in a way that is helpful or cooperative in this group, to what extent do people respond appropriately to encourage that behavior?

Key planning question: How should we respond to one another to encourage cooperation and discourage unhelpful behaviors?

Example methods: Implementing this core design principle begins with creating a shared intent to respond to behavior, not to ignore it either because doing so risks conflict or because one does not have sufficient time. Beyond that, methods such as creating clear agreements for consequences, buddy systems, and coaching-based conversations about performance can help promote cooperative behaviors and discourage uncooperative behaviors, as we'll explore in more detail in chapter 11.

Core Design Principle 6: Fast and Fair Conflict Resolution

Any group that involves committed individuals acting authentically will inevitably encounter conflict, because people have different interests and information. It is best to plan for conflicts and their resolution from the beginning, even if conflict seems unlikely in the future.

Key values: Respect, acceptance, courage.

Key behaviors: Processes and skills for resolving differences in perspectives and aims.

Key outcomes: Timely resolution or accommodation of differences in ways that honor individual interests and perspectives while preserving group effectiveness.

Key assessment questions: How are conflicts usually resolved in this group, if at all? Are these processes fast and fair?

Key planning question: How should we resolve the inevitable conflicts and differences that will arise within and between groups that are authentic?

Example methods: As we'll explore in more detail in chapter 12, there are many methods that support fast and fair conflict resolution, including the development of listening and assertiveness skills and the creation of a role for trusted impartial mediators, a judicial committee with rotating membership, and an escalation process that goes from self-reflection to one-on-one conversation to mediated conversations or even arbitration.

Principles 7 and 8 deal with the way the whole group relates to other groups, rather than the within-group relations of its members. These principles seek to preserve the values of purpose, belonging, inclusiveness, empowerment, transparency, learning, competence, respect, acceptance, and courage at the level of between-group relations.

 Core Design Principle 7: Authority to Self-Govern (According to Principles 1–6)

The seventh core design principle shifts the focus from the internal social organization of the group toward external relation. Every group is embedded in a larger society that can limit its ability to govern its own affairs. These constraints can interfere with the objectives of the group and the implementation of design principles 1 through 6. For example, state educational policy might impose limitations on how classes can be taught, or a military group may need to get permission through the chain of command before it can act in an emergency situation. To create really high-performing groups, it is essential to provide an environment that does not excessively interfere with their capacity to implement principles 1 through 6.

Key values: Group autonomy.

Key behaviors: The group as a whole takes responsibility for self-management.

Key outcomes: The group has the authority and capability to manage principles 1 through 6 to fulfill its purpose (without excessive interference from other groups with different purposes).

Key assessment question: Does the group have authority to govern itself without excessive interference from outside the group?

Key planning questions: How can we take responsibility for managing our own affairs? How should we lead and how should we protect ourselves from undue influence from outside the group?

Example methods: Sociocracy, Holacracy, and other approaches to creating self-managing teams enshrine a group's authority to self-govern. Groups existing in strong hierarchical structures can sometimes find it almost impossible to break out from excessive interference from other parts of a system. Consider, for example, the ways that arbitrary changes in policy following the news cycle, or an excessively restrictive bureaucracy, restricts the capacity of many government employees to perform their work and manage their teams. In such situations, groups under hierarchical control can sometimes make a case for greater self-management in the interest of higher performance, or enlightened leaders can shift organizational cultures to empower groups and the individuals beneath them in the hierarchy.

Core Design Principle 8: Collaborative Relations with Other Groups (Using Principles 1–7)

Principle 8 can be seen as a metaprinciple that allows all the other principles to be scaled to groups of groups. It concerns the way a group relates to the others with which it is in contact. If we are to build systems of cooperation, a group needs not only to practice principles 1 through 7 itself but also to relate to other groups using these principles. In this fashion, the same design principles are relevant at all levels of a multitier hierarchy of social units. The use of principle 8 can go wrong in two ways: first, other groups may not cooperate with you (for example, they don't include your group in important decisions, behave in ways that can't be monitored, and so on), or, second, your group may not cooperate well with other groups.

Key values: Systemic effectiveness.

Key behaviors: A group cooperates with other groups in a system that embodies principles 1 through 7.

Key outcomes: Group relations throughout the system of interest embody principles 1 through 7. This is the concept of polycentric governance we introduced in the previous chapter.

Key assessment questions: Does the group have purposeful, equitable, inclusive, transparent, responsive, harmonious, and autonomy-supportive relations with other groups? Does the group primarily serve its own interests, or those of its larger context?

Key planning questions: How can we have better relations with other groups? How can we contribute to building whole systems that work?

Example methods: As a metaprinciple that includes the first seven principles at any hierarchical level, this principle does not have methods of its own. However, when considered at the level of groups of groups, this principle is deemed to be working well when a group has systems for reporting to other groups, or for coordinating between groups in a purposeful way. For example, with sociocracy, every subgroup is related to every other subgroup through two roles that ensure the interests of both groups are represented. This "double linking" goes a long way toward encouraging purposeful, equitable, inclusive, transparent, responsive, harmonious, and autonomy-supportive relations between groups.

The Core Design Principles as a Multilevel Framework

Having introduced the idea that social systems are continuously evolving at multiple levels (chapter 1) and the core design principles (chapters 2 and 3) themselves, we want to pause for a moment to give you a couple of examples of how thinking from the perspective of the commons, and applying the Prosocial principles at multiple levels of a system, can make a practical difference in improving collaboration on issues that matter.

In cities throughout the world, there are often numerous government agencies and not-for-profit and for-profit organizations working to assist disadvantaged peoples. As just one example, in Australia many towns in the Northern Territory have dozens of different groups working to improve the welfare of indigenous Australians. These groups often compete with one another for resources, so they have little reason to share critical information with one another or to provide support for the work of others. This lack of cooperation results in increased costs and reduced benefits for indigenous people. Without a framework for thinking about how to help not just groups, but also groups of groups, to cooperate, the groups work at cross purposes despite their best intentions. Efforts to create cooperation through top-down regulation or market incentives have all failed, and, at the time of writing, Paul is exploring working with the Northern Territory Police Force to implement the Prosocial process to enhance cooperation among these groups.

Another example comes from Australia's capital city, Canberra. The Canberra Innovation Network is a government-funded agency designed to create an "innovation ecosystem" in which entrepreneurs and start-up ventures are connected with universities, venture capitalists, business strategists, government agencies, and other organizations and individuals. It provides training and assistance, as well as community events to foster more collaborative relationships. The idea is to fuel social and economic growth and have Canberra recognized as a clever and creative city around the world.

But getting this ecosystem to work has not been easy. All the entities in the system compete with one another for market share and recognition. This competition has shown up in selfish behaviors, such as people withholding information that might help the whole network grow or representing only their own interests when making overseas visits to trade fairs and other events that might potentially benefit the entire ecosystem.

So, what might one do to build collaboration in this context? You could regulate to try to force the entities to represent one another's interests and to collaborate. However, this would largely be ineffective and would most likely increase resentment—you can't force people to play well together. Or perhaps you could create market-based incentives for collaboration. While this might have more of a chance of succeeding (since it does at least provide a carrot rather than a stick), it is highly likely that entities in the ecosystem would look for ways to gain incentives with a minimum amount of effort and commitment, and as soon as the incentives were removed, the entities would go back to their old ways.

The Prosocial process and the idea of the commons provided an alternative approach. The principles we just described apply just as well to relationships between organizations as they do to relationships between individuals. Using the techniques we present throughout the remainder of this book, Paul was able to begin a conversation among the groups within the ecosystem to create more shared identity and purpose, equity, inclusiveness in decision making, transparency, effective patterns of mutual reinforcement, conflict management, and authority relations *between groups*.

Auxiliary Design Principles and Exceptions to the Rules

These eight design principles are called "core" because they are needed for nearly all forms of cooperation. Groups can use them to construct the conditions in which altruism can thrive. Monitoring, sanctions, and effective conflict resolution make it safe for us to call out and seek to resolve uncooperative behaviors. Inclusive decision making protects against decisions that benefit a few at the expense of others in the group. Proportionality of benefits to contributions incentivizes effort and discourages free-riding behaviors. Shared purpose and identity create the very ground for collective action. And being able to do all this without excessive interference from outside protects the group from inappropriate constraints that are not locally adapted, while also shoring up the identity of the group.

These principles are necessary but may not be sufficient for any particular group. The Prosocial process leaves space for groups to invent auxiliary design principles that might be just as important for particular groups as the core design principles. As one example, a college fraternity must be structured in such a way that takes into account the high turnover of its members. A classroom might need to be structured to make long-term learning objectives rewarding over the short term. Although we focus this book on the core design principles, which we think are relevant for all groups, we do not wish to restrict groups from creating their own design principles that are relevant to their own context.

Also, while the core design principles are generally useful for all groups, in some situations some of the principles will be less salient. For example, family groups that can rely on kinship relations to create shared purpose and identity may not need to rely as heavily on some of the other principles to maintain individual commitment to the group. Similarly, in dire emergency situations, such as immediate responses to floods or hurricanes, there may be little time for issues such as inequity or inclusiveness to become relevant. In those situations, the very clear sense of purpose—surviving—is enough to maintain prosociality. All this to say that it's important to be sensitive to context, even when using the core design principles!

Seeing the Principles as a Whole

As we stressed in the previous chapter, none of these principles is particularly new, and, in our experience, many groups *know* about the principles, at least intuitively. But many groups don't *follow* these principles, and few of us think of them as a collection of principles that must be implemented together.

We have seen groups go wrong in many different ways. Some groups see relating as a kind of technical challenge. From this perspective, if the group can just design the right rules, procedures, and structures, cooperation will naturally follow. This view of groups as a kind of machine with cogs to be designed into place doesn't survive the reality of our evolved human biases and frailties. For this reason we include evidence and techniques from psychology (see chapters 4 and 5) to help you flexibly adopt the principles. These techniques help humanize conversations so that people are engaged as whole human beings, not as passive cogs in a designed machine.

Other groups seem to lurch the other way, focusing almost entirely on creating positive relationships, emphasizing harmony, inclusion, and equity while ignoring the possibilities of effective design. But these groups place far too much emphasis on individual relations, and they seem ill prepared for the stark realities of conflict and for people or groups acting out of self-interest. In these groups, attending to the capacity to create agreements with backbone is enormously helpful.

Some groups pay almost no attention to fostering within-group relations, and their entire focus is to look good to other groups, such as senior managers or other stakeholders. Other groups focus too much on internal mechanisms and issues and ignore the broader strategic environment in which they operate.

In part, these difficulties of balance develop because of the lack of an overarching framework that highlights the importance of shared purpose and identity, mechanisms to discourage individualistic responding and to encourage cooperative responding, and the ability to move flexibly between groups of individuals and groups of groups. And, in part, groups experience such difficulties because the principles are relatively easy to describe but much harder to implement given the complex reality of human social psychology.

Changing and moving in a valued direction is hard, as we are reminded every year when we first make and then fail to achieve most of our new year's resolutions. Without commitment and action at the level of individuals, values-directed change is unlikely for groups. In the next two chapters we'll describe the psychology of helping individuals and groups flexibly move in valued directions, and then the specific methods we use to create that change.

Practice

The best way to get familiar with the core design principles of the Prosocial process is by applying them to a group that you care about. It might be a work group, your family, or a community-based organization. It doesn't matter as long as it's a group that is important to you in some way.

Using the following diagram, rate how effectively your group implements each of the principles. If the group implements the principle very poorly, place a cross near the center of the circle. And if you think it implements the principle very well, place a cross near the outside of the circle. Once you join all the crosses you'll have a "radar" diagram that depicts how well your group is functioning.

If you want to do a similar exercise with your whole group, you can readily access online surveys at our website (http://www.prosocial.world) that will allow you to collect data anonymously for an entire group or group of groups, and then compile that data into a useful report you can discuss as one of your first steps toward improving your group.

EXERCISE: Do a spoke diagram for a group you care about.

Rate how well your group implements each of these principles (very poor = toward center, very good = toward edge).

Join the ratings to create a core design principle wheel.

48 Prosocial

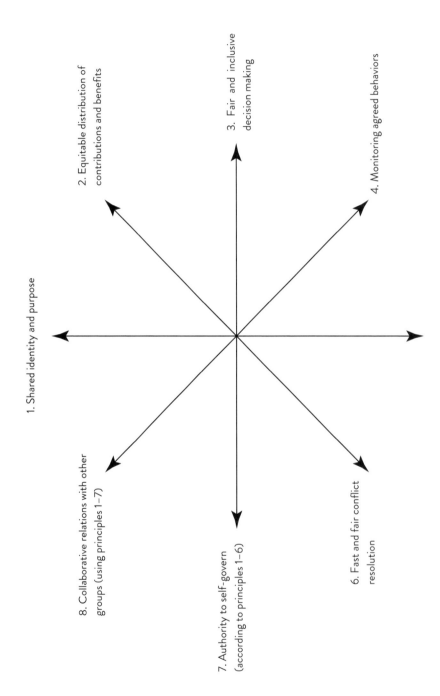

CHAPTER 4

Evolving Behavior
A Contextual Behavioral Approach

So far we've explored how evolutionary theory provides us a way to generalize Ostrom's core design principles to all groups. And you've learned the principles themselves. So now you should just be able to implement the principles, right?

Not so fast. The principles are very general, which makes them very applicable to all groups and also scalable to groups of groups. But our experience has been that just learning about the principles intellectually leads groups to adopt them as rigid rules rather than flexible guides to improving group cooperation. For example, when the principles of monitoring (principle 4) and responding (principle 5) are implemented as a rigid set of performance-management techniques based on top-down measurement and coercive control, they can be as unpleasant as they are unhelpful. We need to learn how to implement the principles flexibly.

Working out the specifics of how the principles might be applied in any given group involves conversations, and those conversations can be very challenging, as issues such as purpose, equity, and conflict can raise strong fears, judgments, and past resentments. Groups need tools to help them navigate these waters without stumbling into entangling conflict on the one hand or avoiding all the hard issues on the other.

The principles point to the direction groups need to move, but psychological skills are needed to make that journey boldly and flexibly. In this chapter we'll begin to lay out the knowledge base you'll need to understand the psychological skills that are required for using the Prosocial process.

Psychological Change as an Evolutionary Challenge

It is easy to get professionals who work in behavior change to quarrel about whether behavior is best understood as an individual phenomenon or a social one. In most areas of study, psychologists are rather individualistic, and you can catch even great minds saying things like, "Groups don't behave! People do! And people are *individuals*," thereby ignoring that every aspect of our meaning making is inherently social. Disciplines that are more group oriented make the opposite error by claiming that all behavior is social, and that a focus on the individual is useless, thereby ignoring the possibilities for individuals to interpret and shape their realities.

The deep lesson of multilevel selection is that evolution occurs at multiple levels at the same time, and thus behavior change needs to be thought of at multiple levels simultaneously. It is both/and, not either/or. That might sound complicated, but we all manage it, all the time.

Balancing our needs with others is the warp and weft of daily life. Will I volunteer to help on someone else's project? Will I interrupt someone who's been speaking too long or wait until they have finished? Should I pick up my dog's poop? What should we do about refugees wanting to enter our country?

Day-to-day human life is an ongoing "social dilemma"—a creative tension between our own interests and the interests of others. If you take a brief moment to think back on your day right now, you'll see situations containing at least a little tension between your own interests and those of others. The closer you look, the more tension you'll see. We live inside the both/and of individual and social life. We live inside multilevel selection.

Elinor Ostrom was awarded the Nobel Prize for work showing that the core design principles accounted for group success, not for work showing how to get people to adopt these principles, nor for work showing how to get them to adopt the principles in a flexible way. We need principles of behavior change to do that work, and those principles are in turn built on the science of learning.

As with our earlier chapter on the principles of evolution, the remainder of this chapter is intended to provide the conceptual basis for a more flexible, practical understanding of the Prosocial process. We begin by reviewing basic processes of learning that culminate in the uniquely human ability to create and understand symbols. We then explore how people can develop a new capability called psychological flexibility, which allows us to act in ways that move us in the direction of our values even in the presence of difficult thoughts and feelings. We explain how psychological flexibility can be understood as an evolutionary process in its own right, involving the enhanced variation, selection, and retention of behaviors in context.

This chapter is for those interested in having a deeper understanding of the psychological processes and evidence underpinning the way we think about learning and behavior change. However, we recognize that some readers will want to move directly to what to do with their groups. If that describes you, you are welcome to skim through this chapter before returning to detailed reading in chapter 5, where we show how you can apply the psychological flexibility model in the context of groups using a specific and easy-to-understand tool called the matrix.

Evolutionary Principles of Behavior Change: Ancient Learning Processes

In the Prosocial process we are seeking to change behavior to create greater cooperation in groups. For the purposes of this book, we can think of "learning" and "behavior change" as synonymous. Learning (that is, behavior change) in humans is built upon layers of different forms of learning acquired by organisms through evolutionary history. Because our bodies are built upon the foundations of that evolutionary history, we still make use of all these forms of learning, and all of them affect behavior in groups. At the same time, as cooperation is affected by our most sophisticated verbal reasoning, it is also affected by emotional systems originally developed hundreds of millions of years ago in prehuman organisms.

While we could have written this book without describing the nature of learning and behavior change, we think something would have been lost. Most people think about evolution in purely physical terms, focusing on genetic evolution. But as we described in chapter 1, the capacity to learn is in fact itself a whole new stream of evolution. Whether we create greater global cooperation or greater global enmity, and whether we preserve the biological systems of our planet or destroy them, depends upon our capacity to learn and change our behaviors individually and collectively. Learning about learning and behavior change helps make you more flexible in your implementation of the Prosocial process—able to spot and respond more appropriately to more "primitive" forms of reactivity in flight and able to adapt and develop new techniques for working with groups to fulfill the underlying intentions of the Prosocial process. We find that many people using the Prosocial process want to understand why it works so that they can more creatively apply its methods while remaining true to the underlying principles. If that describes you, let's get into it.

The oldest (600 million years old or more) form of learning (that is, behavior change) is habituation, in which repeated irrelevant sensory stimulation leads to a reduction in responding.[1] We all show habituation when our startle response decreases dramatically from the first loud bang to the last during a fireworks display. There is little we can do to intentionally work with this form of learning in groups. It happens automatically.

During the Cambrian Period a bit over half a billion years ago, animals developed the capacity to associate events that preceded and reliably predicted other, later events—for instance, rustling sounds that reliably predicted the arrival of certain predators. This kind of learning (called classical conditioning) prepares the organism to respond more rapidly and effectively to previously experienced regularities—in effect getting a jump on what experience has shown is likely to come next. For example, when you go to the dinner table it's useful that you are already beginning to salivate because it means you are more ready to eat. An animal responding emotionally to rustling sounds that were previously associated with a predator is better prepared to survive by running, fighting, or freezing.

In the context of human groups, you can see this sort of learning in play when, for example, a person reacts unreasonably emotionally to a perceived threat. It is possible that some learned association from similar events in the person's learning history, perhaps as a child, has triggered them. Victims of post-traumatic stress often demonstrate this sort of stimulus-driven reactivity, but it isn't necessarily dramatic. A notable portion of our emotional responding involves an automatic response to events that are similar to those in our history.

Soon after classical conditioning evolved, some social animals also developed forms of social learning, such as mimicry, requesting social guidance, or imitation. We see this operating in humans when a crowd begins to clap in synchrony, for example, without any conscious intent.

A particularly important and more complex form of learning also developed during the Cambrian Period, when animals evolved the capacity to detect and respond not just to antecedent conditions (as in the forms of learning described thus far), but also to regularities in antecedent-action-consequence sequences—what are called contingencies. In a given motivational and situational circumstance, organisms tend to repeat behaviors that result in relatively positive outcomes and avoid behaviors that result in relatively negative ones. (We mentioned this in chapter 1 when we described how a Skinner box works.) This kind of learning (called operant conditioning because it strengthens actions that "operate on" or change the environment in predictable ways) is essentially a fast-paced process of behavioral evolution built by the slow-paced process of genetic evolution.[2]

Learning principles especially provide the beginning of a very useful theory of behavior that we can use to promote intentional change within and between groups ("behavior" here means anything and everything people can do even if no one else can observe it directly, such as remembering). Given an *antecedent* context, particular variations in *behavior* produce particular *consequences* that select for that behavior in a particular *motivational* context. For example, if we want to do something about a group member who seems resistant to change, maybe instead of just pushing harder (a change to the antecedent social context) we should ask if the consequences of change are likely to be better or worse for them compared to their current situation (a change to focus on the consequences of behavior). If the consequences are relatively negative, it's not surprising that the action is less likely to reoccur (figure 4.1).

Operant behavior is inherently purposeful, but in a limited sense of the term. The animal acts "in order to" produce consequences that have been secured in the past by various forms of action. It is not the actual future that controls this kind of learning—it is the "futures" that have been experienced previously.

Operant learning is functional. What is reinforcing for a particular action, of a particular person, in a particular time, situation, and motivational state, may not be so for another action, person, time, situation, or state. To apply operant learning we must ask: what matters and why? Figure 4.1 offers a brief schematic of this way of thinking. To better understand behavior it helps to pay attention to both the antecedent context that people are responding to, as well as the history of consequences they have experienced for their behavior. Once a behavior has occurred, the consequences form a new antecedent context for their own and others' behavior. This is why we added a feedback loop to the figure.

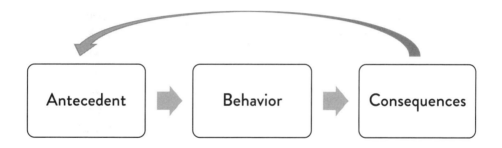

Figure 4.1. Behavior occurs in response to a context, and it produces consequences that make similar behavior more or less likely in the future.

If we want to build a science of intentional change, it makes sense to use the power of associative, social, and contingency learning to affect behavior. We cannot stop there, however, because these tools do not yet fully give us the means to foster key elements of prosocial relations to others, such as trust, awareness of shared purpose, and more prosocially oriented values. For that purpose, we need also to delve into the world of narratives and rules, cognitive heuristics and problem solving, and sense of self and perspective taking. In other words, for that purpose we need principles of symbolic learning.

Evolutionary Principles of Behavior Change: Symbolic Learning Processes

In developing the Prosocial program we turned to methods drawn from contextual behavioral science, in part because of its well-developed research program on symbolic behavior, which helps facilitate the adoption of the core design principles in groups.

Symbolic learning differs from normal associative and contingency learning because it is built from the bottom up on the two-way street of *relations*. A normally developing twelve- to fifteen-month-old baby who learns that a round rubber thing is a "ball" will know without specific training to look for the round thing when later hearing that word. Names are not "symbols" for objects unless this kind of two-way street emerges, in which learning a relation in one direction (round thing that bounces = ball) means you can derive the corresponding relation in the other direction (ball = round thing that bounces) and then combine those relations into symbolic networks. For example, a child learning that another name for ball is "palla" (as it is in Italy) will look for the object when hearing "palla," even if the word was never said in the presence of a ball.

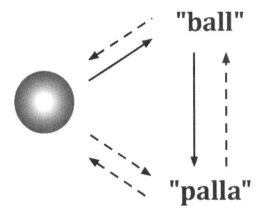

Figure 4.2. Directly trained relations are shown by solid arrows; derived two-way, or combined, relations are shown by dashed arrows.

The important takeaway from figure 4.2 is that the dashed lines representing derived relations do not have to be trained one by one once the child knows how to derive symbolic relations.[3] They come along, as it were, for free. We take this ability for granted because we do it so easily, but research has yet to reliably show that any other species does this kind of two-way relating.

Relational learning doesn't happen just with names. Consider any verbal relation and you will note that they have a two-way quality that can then enter into networks. If Sally "is bigger than" Sam, you can derive that it is equally true that Sam "is smaller than" Sally. If you later learned that George was bigger than Sally, you could derive the relative size of all three.

To appreciate the increase in variation afforded by relational learning, consider eight objects that are related in a particular way to eight symbols (this is called that, this contains that, this is bigger than that, and so forth). If this example were taught to a nonhuman animal, you'd have only the eight trained relations. But humans can derive many more. If all events, all relations, and all symbols are related to all other events, relations, and symbols in all possible ways, how many relations can be derived from these sixteen events with their eight trained relations?

About four thousand.

Given this capacity to derive relationships, how many possible relations could you derive between events inside your own mind? The answer might as well be "as many stars as there are in the sky." The answer is incalculable.[4]

This combinatorial property of relational networks is not only massively generative, it also changes how we solve problems, such as how to get along in groups. Verbal problem solving requires a symbolic description of events and their attributes entering into an if/then relationship with imagined outcomes that can be mentally compared: "Given this situation and its properties, if I do this, I will get better outcomes than if I do that." In normal operant learning, the future purpose of action has actually been experienced in the past; in symbolic learning, the past controls the cognitive construction of a purpose that may never before have been experienced.

This new form of purpose (one based on symbolic construction instead of direct experience for the individual) creates a new contingency stream for evolution—not just contingencies of reinforcement, but also contingencies of meaning. Instead of just relying upon automatic reactions, thanks to the combinatorial property of relational networks we can now plan ahead and create purpose symbolically.

The symbolic mind is a game changer for life in groups, and it has led to our greatest successes as a species, including science, technology, art, and literature. However, symbolic language is also a double-edged sword—a source of enormous variation but also enormous rigidity. The same symbolic relations that allow us to *solve* problems mentally also allow us to *create* problems mentally.

Imagining positive futures that have never been and mentally comparing the steps that might give rise to them is enormously supportive of human progress. But people also use the exact same skills to imagine frightening worlds that also have never been, and feel overwhelmed or incapacitated as a result.

A physician can detect sources of pain and work to eliminate them, but when people apply symbolic problem-solving rules to try to avoid such sources of pain as unpleasant memories or self-critical thoughts, it often doesn't help them personally or socially. For example, a person who withdraws from a group to avoid thinking about a painful social rejection is at risk of entering into a self-amplifying loop in which avoiding the pain of rejection and loneliness produces actions that ensure further withdrawal and loneliness.

The downside of human cognition is a kind of evolutionary mismatch—that is, evolved responses that worked for earlier forms of learning do not necessarily work with symbolic learning. For example, all animals will avoid not just direct sources of pain but also sources of pain based on associative learning (such as the rustling leaves example). That makes perfect sense, and half a billion years is plenty of time to get good at it.

When we formulate sources of psychological pain symbolically—for example, with things like negative emotions or negative thoughts—we also then try to avoid them. This behavior is far more recent in our evolution as a species. Complex human emotions and patterns of thought did not *even have names* in the early years of writing, and written language is only about five thousand years old. In fact, some of our "emotions" (for example, depression) have only had names for a few hundred years.

Using problem-solving rules to figure out how to avoid things like depression or a lack of confidence is logical, but it can easily lead to psychopathology, in part because symbolic learning processes can become self-contradictory and self-amplifying. For example, a problem-solving rule like "I have to fix my car or tomorrow will be terrible" is rational and useful. "I have to get rid of my anxiety or tomorrow will be terrible" seems to be the same but is likely to cause harm because imminent terrible events *generate* anxiety, and in effect this rule means that anxiety is now something to be anxious about. Self-amplifying processes of this kind can easily go awry.

This same double-edged quality applies to cognition in groups. When narratives and heuristics are collectively held, they drive the behavior of human groups. We occupy a shared symbolic world of meaning systems in the groups to which we belong. Once again that is both positive and negative. A shared meaning system allows us to tell stories that articulate group purposes and agreements for how we should behave, but they can also be used to exclude and attack others with differences in culture, ethnicity, religion, language, or a host of similar marks of group identity.

There are other examples of symbolic learning working against cooperation. Symbolic rules can be applied in ways that are coercive or that pull for excessive social compliance. Agreements that "this is how we do it here" or narratives about the supposed skill of the group ("We always win") can blind the entire group to healthy alternatives or to changing contexts that require new group practices. Members of a group can hold tightly to symbolic labels about who they are or what their role is in the group, which can induce behavioral rigidity (for example, "My boss did me wrong and I am going to find a way to get back at her"; "Joe is evil"; "I am just an introvert—I'm not good at being a leader").

In summary, symbolic learning is a very recent form of learning that has completely transformed the lived experience of being human. Once we become verbal, we spend the rest of our life evaluating, comparing, naming, wishing, and imagining. Technologies from the printing press to the Internet and smartphone have put this form of behavioral and cultural evolution into overdrive so that our lives are increasingly "virtual" and disconnected from the direct physical contingencies of existence.[5]

Psychological Flexibility

Because symbolic (that is, relational) learning is key to our development as humans but leads to both positive and negative effects everywhere we look, a primary task is to put this kind of learning on a leash so that it can be used in a more thoughtful fashion in applied programs of behavior change. People need to be able to deploy their cognitive skills flexibly—in ways that fit the situation and the purposes of action.

Learning how to do this is central to psychological flexibility, which involves consciously moving in the direction of values even in the presence of difficult thoughts and feelings. Psychological flexibility is improved by several methods of psychological intervention, but it was first deliberately targeted by acceptance and commitment training (or ACT, said as a single word, not its initials).[6] The Prosocial process includes several ACT methods that are meant to help motivate individual members to participate and learn to cooperate, both in the Prosocial process and in the work their groups do, and to reduce the barriers produced by symbolic processes that might be holding them back.

Multilevel selection (MLS) helps us think about the double-edged sword of human symbolic behavior and how it might be addressed in a way that comports with psychological flexibility. If you look at life from the perspective of MLS, it is groups all the way down. What we think of as the "individual" is composed of a vast number of entities: organs, cells, specific behavioral patterns, and so on. And in the same way, the "individual" participates in a vast array of nested groups: a family, a neighborhood, a city, a country, the world.

The term "individual" comes from two Latin roots: *in*, meaning "not," and *divide*, meaning "divide." Individuals, the Romans supposed, cannot be divided; they are "indivisible."

That is clearly not true. How we interact at a party, for example, can be completely different from how we act at home. The voices within us from the people we know well have, in a sense, a life of their own. One way of feeling may seem to be impossible to combine with another way of feeling. We are many. That observation is key to understanding psychological flexibility.

Recall from chapter 1 that adaptation at any level of a multitier hierarchy requires a process of selection at that level—and it tends to be undermined by selection at lower levels. Although the "individual" person is indeed one level of analysis, selection also applies to smaller units within. The human body is made up of trillions of cells, any one of which can go rogue and use bodily resources to multiply without regard for the whole. Without the immune system to disrupt this process, cancerous cells would rapidly kill us all. In much the same way, we face selfishness from "groups within" when emotions or cognitions claim an unfair share of time and attention from us.

What psychological flexibility does is provide the skills and perspective needed to rein in internal selfishness that develops from the various thoughts, feelings, sensations, and memories that each of us contains. This more cooperative psychological stance helps us choose better potential actions, so we can best fit them to our current context. In effect, we shift from what's in the interests of parts of ourselves, such as our anxiety or suspicion, to what's in the interests of our whole person. In the Prosocial process, we first cultivate psychological flexibility skills within the individuals in the group. These skills are then applied to the group itself, and, finally, they become the foundation for exploring the core design principles. We depict these ideas in figure 4.3.

1. Within groups, people exclusively focused on their own needs and desires can be disruptive.

2. Broadening perspective makes it possible to discover empathy, shared values, and, ultimately, cooperation.

3. Groups can also focus on their own identity and purposes, ignoring others (for example, a political party that wishes to win at all costs, or a terrorist cell).

4. Again, broadening perspective makes it possible to discover empathy, shared values, and, ultimately, cooperation.

5. And, when it is groups all the way down, behaviors (such as feeling fear) can act selfishly and dominate the psyche.

6. And, once again, broadening perspective makes it possible to discover self-compassion, authenticity, and, ultimately, equanimity with the whole of experience in time ("This too shall change") and space ("What else is going on?").

Figure 4.3

In the remainder of this chapter we'll describe psychological flexibility concepts in steps that parallel how these ideas are actually deployed in the Prosocial process. We will deal first with how our internal choices about what brings meaning to our actions can select our actions. Then we'll look at how we can encourage healthy sources of variation internally and externally, and examine how we can retain successful actions in a way that consciously fits to context.

Selection Criteria for Behavioral Evolution: Values

Nonhuman organisms are more sensitive to immediate reinforcement rather than delayed. In the vast majority of life situations, the rule is "smaller sooner trumps larger later." Even if a delayed reward is quite a bit bigger, actions that lead to smaller and more immediate consequences will generally prevail. That said, this rule is not always true, and it is especially not always true with people. The evolutionarily more recent tool of symbolic learning provides us a way of bringing delayed consequences into the present moment.

To appreciate how this works, let's begin with an important study showing that symbolic learning can be used to harness and modify classical conditioning. Imagine that you are shown three letters repeatedly in random order: X, Y, and Z. Every time Y appears you are given a small electrical shock. How would you respond when you later saw X, Y, or Z? In this example, only Y would likely produce much of a reaction. But what if *before* receiving the shock you were told that "X is smaller than Y, which is smaller than Z"? In this example X still wouldn't bother you much (after all, it's smaller than the letter that led to the shock), but Z may be *very* frightening. When you see Z your heart rate will change, and you will start to sweat. We know this for a fact, because this study was actually done.[7]

What this study shows is that symbolic learning changes how events impact us. That includes imagined consequences of action. When advertisers describe how food will taste, or how a car will drive, or how people will respond when they see you in new clothes, the described consequences become psychologically present, and that can motivate behavior that might produce them, even if the consequences will be delayed or in fact don't occur later. Our purposes grow largely from our capacity to tell stories about the future (and the past).

When we say "purpose," we generally mean either values or goals. *Values* are our chosen qualities of being and doing. They express how we want to be as people, and the desired qualities we would like our actions to reflect. Think of them as the adverbs of our lives. When we act in service of values, we are acting lovingly, compassionately, kindly, consciously, and so on.

Values are different than goals. Concrete goals can be achieved; you can make a million dollars, or get a degree, but once a goal is achieved it will likely lose its power as a source of further motivation. That is not true of values, and this difference is a major reason why values are so important. Values are only displayed or instantiated, not obtained or possessed, so while we may use them, they are never used up. They are a continuously available source of meaning and motivation, no matter what situation we find ourselves in.

METHODS FOR DISCOVERING AND CLARIFYING VALUES

To clarify your values, you might explore the times in your life that were especially sweet—the times when you felt especially alive, uplifted, empowered, connected, or vitalized in a given domain, such as by being part of a particular group. Inside those sweet memories, personal values will likely be revealed.

Conversely, you can take the things that hurt the most and flip them over to see what they say about the qualities of being and doing that you want in your life. For example, if it hurt when your supervisor lied to you, it suggests that honest work relationships are part of what you care about.

You could review your life as a story and write or think about the themes you hope your life will reflect going forward. For example, you might design your own ideal retirement party, complete with the speeches about what your behavior from now forward reflects in a values-consistent world of your own making.

You could also describe your heroes. We are social beings, and people who inspire us can also help us find our own aspirations. Whom do you deeply admire, and why? Who is important to you? Our heroes and important people stand for something, and that is part of why we cherish them so. The things they stand or stood for are generally qualities that we seek in ourselves as well. They reflect our values.[8]

Even thinking about values is itself a values-based thing to do. As soon as you make a choice, the motivational landscape changes. A person who realizes she highly values being authentic will approach situations differently thereafter. Although in the short term it might be frightening to be more forthcoming, this person's desire to be consistent with her values provides a powerful motivation to do so. In effect, her value becomes a selection criterion for behavior. The choice between "I won't feel as frightened, but I'll know I'm being inauthentic yet again" and "I may feel frightened, but I'll be learning how to be truer to myself" will stand out and make finding ways to take the latter path easier.

Asking people what they deeply care about or want to move toward tends to foster a collaborative conversation that can motivate participation and connection with the group, which is why it's one of the first issues addressed in the Prosocial process, as we'll see in the next chapter. Furthermore, if values are shared with others they can begin to provide a selection criterion for the actions of the group as well as of the individual.

Variation: Creating Space for Thoughts and Feelings with Emotional and Cognitive Flexibility

You might think that people would be examining how their actions fit with their values all the time, given how simple and effective it is for creating a life well lived. They don't, and there's a reason for it: other parts of human psychology get in the way—ironically exacerbated by the very symbolic learning processes that make values work possible.

Values uplift us but also make us vulnerable in the word's original sense of "wound-able." In the past, we've been hurt in areas we most care about. That is why both sweet and painful memories can be used as guides to values. If we care about trust, we care both because trusting relationships have been vitalizing and because betrayal of trust has been heartrending.

When we apply our problem-solving skills to emotional pain, we begin to evaluate those emotions. Given only a choice between "good" and "bad," what comes up in your mind as you evaluate these emotions? (Be sure to pause for a few seconds after reading each word.)

Happy

Sad

Anxious

Joyous

"Sad" and "anxious" are usually quickly sorted into the "bad" bin, and "happy" and "joyous" into the good bin. Evaluations like that are built using some of the same skills we use for choosing values (this is obvious in the very etymology of the two words "values" and "evaluation"). If some emotions, memories, urges, or thoughts are judged as undesirable, the seemingly logical thing to do is to remove them as one might remove an external object.

In many cases that approach tends to backfire. What if you are sad because your mother died? Is that kind of sadness bad too, even if it's a natural expression of your love for your mother? What if you are anxious because your work environment is dysfunctional? Is that anxious feeling "bad," even if being bothered by how your work team is functioning is key to you doing something about it?

The tendency for people to do the logical thing with emotions or thoughts (to try to get rid of the "bad" ones and cling to the "good" ones) is unwise. Trying to subtract reactions is unhelpful because there is no delete button in the human nervous system—no psychological process called "unlearning." You can add to a learning history, but you cannot subtract from it. Even things that are forgotten can be relearned again more readily because the organism is changed permanently in some ways by the initial learning experience. If you try to get rid of your painful memories, the memories remain, but as a result of your efforts there are now new avenues leading to them, and more moments connected to them. That is why a suppressed thought increases in frequency[9] as distractors soon function as reminders.[10] The memory and the pain it contains will be more, not less, impactful. And trying to cling to positive relations can be unwise if the effort to do so makes one insensitive to changes in the internal and external environments.

Experiential avoidance, the tendency to try to change the form or frequency of private experiences—even when it is behaviorally harmful to do so—is one of the single-most harmful and repertoire-narrowing psychological processes known, including in group settings.[11] A healthier approach is to add to our history with new and positive experiences, and to change the actions taken in response to difficult internal experiences.

Values-based actions in groups add positive moments to our lives, especially as group development is guided by the core design principles. The principles help you set up a social context for trust, and emotional openness sets up a psychological context for the vulnerability that comes from allowing trust to form. In other words, emotional openness (the willingness to experience emotions without needless defense) empowers group participation.

Let's suppose we have a group member who has thoughts like *Oh, I can't do that because I'm not good at it*, or *It makes me too anxious to call volunteers. Find somebody else.* If cognitive networks work by addition, the person will go to her grave with a capacity to think such things. But thoughts don't always have an automatic behavioral impact, such as leading a person to decline opportunities for participation. For example, noticing thoughts with dispassionate curiosity dramatically reduces their automatic behavioral impact, such that a person might have the thought *It makes me too anxious to call volunteers* and still agree to do so.[12] This effect is sometimes called cognitive defusion, or cognitive flexibility. It has a profound impact on the automatic ways that thoughts can lead to action.[13]

Much as epigenetic processes alter how genotypes are expressed in the production of proteins that alter phenotypes, openness and awareness function as episymbolic processes that alter how our cognitive networks—our symbotypes, if you will—are expressed in our actions, altering how we approach the challenges we face. We can have the thought that we *must* do something (for example, tell off a coworker) and respectfully decline that mental invitation; we can have the thought that we cannot do something (for example, give an important talk) and do it anyway.

Retention: The Practice and Pattern of Committed Action

Although values are an internal symbolic event, values-based overt action is where the rubber meets the road. The benefit of dealing differently with internal barriers is that values can be put into practice through committed action.

Behavioral retention is fostered by two primary features of action: repetition leading to successful outcomes, and the integration of action into larger and more elaborate patterns that pay off in expressing values. As a shorthand, the prime tool for behavioral retention is practice and pattern of values-based action.

Behavior change is hard. So people need to start small with new habits in contexts that are supportive and then repeat them over and over. A person wanting to learn to speak up more in meetings at work might need to start with smaller meetings, or with topics that are not too emotionally difficult. Because success is also important, especially when beginning new habits, it's important to create a positive social context for new actions, such as by ensuring that there is an opportunity for needed training, or that the project or action will be seen as helpful by the group. Behavior change goals should be clear so that progress can be noted.

Patterns of action are also important. As a particular behavior change becomes more integrated with other actions in many different situations, it's more likely that it will be sustained. For example, a habit of being actively supportive of others in the group becomes stronger when it involves a variety of forms of support occurring in a number of areas. Building retention via practice and pattern only works to produce more prosociality if these actions are supported by movement in a values-based direction.

That is what commitment is about. Committed action means pursuing larger and larger habits of values-based action. It means being committed to developing a practice and pattern of moving toward what is meaningful.

Context and Consciousness: Who Notices What Now?

So far we've described four of the six flexibility processes in the ACT model—valuing, cognitive flexibility, emotional flexibility, and committed action. The final two are flexible attention to the present moment and perspective taking. Together we can label these "awareness." The major evolutionary impact of awareness is to increase context sensitivity so that the variation and selection of behavior are more likely to fit the external and internal situation, and thus be retained.

The internal and external environments are always literally in the now, but when thoughts emerge it is quite possible to enter into a symbolic world that fixes attention on the past in the form of rumination, or on the future in the form of worry. When this happens, people lose contact with what is actually present. Emotional and cognitive openness helps people notice these attentional shifts, which are often deployed to avoid experiences we don't like, and instead use attentional flexibility to move attention toward values issues that are at stake, and how the social and external environment affords opportunities for positive action.

Applying flexible attentional skills is easier if people learn to step back and notice that they are doing a kind of perspective taking on themselves. It takes children a long time to learn this ability. Researchers studying relational learning have shown that cognitive skills involving person, place, and time—the three relations children learn growing up, in that order—foster the ability to notice. First, they learn *person*, what is "I" versus what is "you." Then they learn place, "here versus there." Then time, "now versus then." As the three repertoires come together, a sense of "I/here/now" emerges. So, too, does an ability to imagine how things look to others from behind their eyes—their perspectives. This perspective-taking ability helps us "decenter" from our self-interested narratives and empowers us to think of others' needs, not just our own.

When relating to others, active perspective taking has profound social effects when it's combined with the other flexibility skills. Combined, these skills constitute greater mindfulness—a term you've probably heard of that represents open and nonjudgmental attention to the present moment. Mindfulness is the very essence of learning to flexibly and deliberately move between the actual physical world of direct contingencies and the symbolic world of our interpretations and meanings.

A vast amount of literature shows that mindful awareness fosters both our own mental health and our effectiveness and ability to relate to others. Defusion and cognitive flexibility skills at the personal level soften a person's judgments about others at the social level. Emotional openness and acceptance skills focused on the self foster compassion for others. Attention to the now helps us see what is possible in social situations, and perspective-taking skills help us see the world through others' eyes. Flexibility skills help us create a platform of mindful awareness so that we don't privilege our own needs over those of others, thus creating space to act more prosocially.

How Psychological Flexibility Fosters Key Aspects of Cooperation

We've covered the six flexibility processes that make up psychological flexibility: values, emotional flexibility, cognitive flexibility, committed action, flexible attention, and perspective taking. Together these comprise an integrated skill set. Both individuals and groups can be psychologically flexible. It is a skill set that supports prosocial behavior change and the deployment of the core design principles. However, it also has implications for building some key elements that social psychologists have found underpinning prosocial behavior: trust, a focus on longer-term rather than immediate outcomes, and social value orientation.

Trust

The first target for behavior change is *a sense of trust*. If you know someone who tends not to trust others—or if you yourself find trusting difficult—then you know how much a lack of trust can interfere with cooperation. Any slightly ambiguous situation can easily be interpreted as an act of aggression or selfishness, setting off the person to act more selfishly themselves, thereby ensuring that cooperation falls apart. When people are able to trust others, not only are they more likely to cooperate, they are also more likely to reciprocate when others cooperate with them. Some of these differences are trait-like, some are accounted for by the structure of groups in which people participate, and some are due to psychological processes that can be changed. Even the most untrusting person will be more likely to trust in some social situations; even the more untrusting soul can become somewhat more trusting over time. On the social side, the Prosocial process is designed to build trust by deploying the core design principles, which create social situations in which vulnerability is less likely to be exploited. On the psychological side, the Prosocial process uses knowledge of the processes underlying both direct and symbolic learning to dampen the internal barriers to trust.

Psychological flexibility fosters trust for a number of reasons. Trust entails a sense of vulnerability. For trust to develop something must be at stake.[14] Emotional and cognitive openness facilitate the sharing of values and vulnerabilities and reduce the pull to avoid. Perspective taking fosters empathy and connection, and when combined with greater psychological openness it predicts more enjoyable and less judgmental social relationships.[15] Perspective taking also helps people to better predict the needs and reactions of others so that they can avoid accidental violations of trust. The sharing of values allows for greater interpersonal connection and transparency about motives, further fostering trust.

Long-Term Thinking

The second target of behavior change is to shift attention to the longer-term implications of actions. People tend to act more prosocially when their attention is less on their immediate reactivity and more on what they want for the longer term.[16] That, too, is trait-like to a degree, but it's also modified both by social context and psychological skills. When social situations are predatory or can change without notice, it pays to be more impulsive. A long-term view requires predictable rules of social engagement, and the development of long-term supportive relationships. It also requires that one has the symbolic skills needed to focus long term, and the psychological skills to sit with the emotional urges to act impulsively without doing so.

Values work and the reduction of impulsivity that comes from greater emotional and cognitive flexibility foster a healthy future focus. Research has shown that cognitive defusion and emotional openness can reduce people's focus on "smaller sooner" consequences over long-term consequences (what is called delay discounting).[17] This makes sense since it is aversive to wait for positive events or to sit with negative ones, and tempting but avoidant solutions to this problem will appear.[18] Conversely, values-based commitments undermine such impulsive processes.[19]

Social Value Orientation

The third and final target for behavior change is developing more of a *social value orientation*: the degree to which you care about others' outcomes relative to your own.[20] Before we say too much about this idea, we want to give you a real taste of what it looks like.

Imagine you've been brought into a lab, seated in front of a computer screen, and told that you have a "partner" called "Other" whom you don't know and will never meet. Both you and Other will be making choices by circling either the letter A, B, or C. Your own choices will make money for you and Other. Likewise, Other's choices will make money for her and for you. The more money you receive, the better for you, and the more money Other receives, the better for her.

If you select...	A	B	C
You get	$5.00	$5.00	$5.50
Other gets	$1.00	$5.00	$3.00

In this example, if you chose A you would receive $5 and Other would receive $1; if you chose B, you would receive $5 and Other $5; and if you chose C, you would receive $5.50 and Other $3. So, you can see that your choice influences both the amount of money you receive and the amount of money Other receives.

Now, for each of the three situations below, circle A, B, or C. Keep in mind that there are no right or wrong answers—choose the option that you, for whatever reason, prefer most. Also, for this task, please imagine that this situation and the money you'll receive is real—you and Other will actually take home the amount of money each of you accumulates.

Situation 1			
If you select...	A	B	C
You get	$4.80	$5.40	$4.80
Other gets	$0.80	$2.80	$4.80
Situation 2			
If you select...	A	B	C
You get	$5.20	$5.20	$5.80
Other gets	$5.20	$1.20	$3.20
Situation 3			
If you select...	A	B	C
You get	$5.60	$4.90	$5.00
Other gets	$3.00	$4.90	$1.00

Now, write down the three choices you made in sequence.

In each situation, there is one choice that maximizes your own earnings (B, C, A, respectively). This is called the individualistic response, and it is closest to the concept of *Homo economicus* (optimize your own outcomes irrespective of others). For each situation, there is also a prosocial choice that maximizes Other's earnings (C, A, and B, respectively). If you chose this option, you are willing to take a small reduction in your own outcomes in order to maximally benefit Other. Finally, in each situation there is a competitive choice that maximizes the gap between how much you get and how much Other gets (A, B, and C, respectively). What did you choose?

A longer version of this task has been used in hundreds of scientific studies. We know that social value orientation is important for prosociality. As with trust and consideration of long-term consequences, social value orientation has a trait-like aspect, a social situational aspect, and a changeable psychological aspect. It is possible to shift people from behaving like individualists and competitors to behaving more prosocially, even if the trait-like part stays the same.

The strategies employed to do this depend on where you start. Research has shown that people who start from the posture of "What's in it for me?" change in a prosocial direction if the personal benefits of cooperation appear to exceed the costs.[21] That process is even visible neurobiologically. When such choices are made, areas of the brain associated with calculating extrinsic benefits (the lateral prefrontal cortex) are activated and then pass on information to the reward system (occupying a region from the striatum to the ventromedial prefrontal cortex).[22] As a practical matter, this means that interventions designed to increase prosociality will sometimes need to be focused on increasing a sense of personal psychological and practical benefits while decreasing personal psychological and practical costs. The psychological tools in the Prosocial process are designed to do just that.

People starting more with the "What's in it for us?" posture change when they can be sure that others in their social context can be trusted (also a process that's visible neurobiologically). This make perfect sense: aiming for a win-win exposes people to the risk of exploitation, and thus paying attention to trust is more important, *especially* when the stakes are high and the risks are greatest. That's why the relationship between trust and cooperation is strongest when there is more rather than less conflict.[23] The psychological tools in the Prosocial process are designed to be helpful here as well.

Strong social agreements create positive expectations of others. We all trust complete strangers with our lives when we drive on the road or fly in an airplane because of the strong network of cultural practices and rules surrounding transport. In social situations, transparency about other people's intentions, the reduction of uncertainty, and the opportunity to impact the social process predict more cooperation.[24] Strong social agreements are part of the reason why the Prosocial process works.

Social values are dependent on perspective taking and a willingness to seek win-win solutions, so becoming more psychologically flexible is potentially a way to enhance social value orientation. The congruence between personal values and social values is a predictor of health.[25] Conversely, psychological inflexibility leads to the objectification and dehumanization of others, as in stigma and prejudice.[26] Getting clear on social values, creating space for even difficult thoughts and feelings, and committing to values-consistent action is part of the Prosocial process for working with people to enhance cooperation.

The following exercises can help you explore your own psychological flexibility.

EXERCISE: Picking a Guide or Hero

To explore the relevance of psychological flexibility to group processes in a more experiential way, consider a group you are part of that you care about but that has some degree of conflict. Maybe it's the group you're hoping to use the Prosocial process with, maybe not. Then, in a general way, allow yourself (without formulating any easy answers, rather just "sit inside" the issues) to try to open up to how it feels to be in the group, to evaluate whether you have a sense of trust in the group, and to determine whether the long-term welfare of the group matters to you. Then, as you immerse yourself in these issues, consider this question: Given what I'm feeling, who in my life might be a good guide to help me determine how to be an effective part of this group, or who do I consider a hero in terms of how they live their values in situations like the one I'm dealing with?

Allow your mind to settle on someone. If your gut sense tells you that the person is an apt choice, trust the process and see where it takes you—even if initially the person seems implausible.

Now that you've picked a guide or hero, ask yourself some questions. Sit with each question for a minute or two and then perhaps write your answers down or speak them aloud. Don't rush through the questions. Pause to think about each one.

- What values does this person represent to you? If you had to say what the person stands for, what would it be?

- Are those values ones that you would like to manifest more in your own behavior?

- If this person could be here looking at you right now, what do you think they would see? Would they see a worthwhile person in you? Would they accept you for who you are? Would they judge you harshly, or not at all?

- If this person were aware of the issues you're having in the group, and they were willing to provide guidance, what do you think your guide would advise?

- What is your gut sense about that advice? Is it wise? Could you take steps to adopt it?

- Finally, before finishing, if you see anything worthwhile in this exercise, please note that this guide is actually "in you" and in your history, and thus they will always be but a moment away. Your guide or hero is with you and will be available to you so long as you are alive.

EXERCISE: Me and Us

Noticing the ebb and flow of acting in the interest of self and of others can help us develop a sensitivity to our individual interests versus those of the group. Here are some steps to help you:

1. Reflect on three times today when you made large or small decisions to help someone else or to cooperate with them. What factors shaped your actions?

2. Looking back over the past month, can you think of a time when your own interests conflicted with those of others? What was this like? What factors influenced your decision making?

3. Try to bring the issues you have just identified into your next day in a more moment-to-moment way, as you naturally work to balance action as a multilevel process.

Conclusion

Multilevel selection reminds us that behavior change in groups cannot be separated from behavior change in individuals, and vice versa. In this chapter, we've explored the concept of psychological flexibility as the conscious evolution of our own awareness. Psychological flexibility establishes new, more workable selection criteria for behavior because individuals and groups work to clarify values. It enhances variation by making more room for thoughts and feelings, and it enhances retention by establishing and maintaining patterns of committed action. Attending to the present moment and perspective taking facilitates an awareness of the context for action. All these processes, taken together, help build trust, consideration of the future, and a prosocial value orientation, thereby enabling more prosocial behavior.

As the following chapters will show, these flexibility skills can be increased fairly simply. Individuals and groups can then use them to foster the adoption of the core design principles. In the next chapter, we'll look at the methods used by the Prosocial process that can help you increase psychological flexibility and the capacity for values-based action in all group members. These methods lay the groundwork needed for group members to adopt the core design principles.

PART 2

Prosocial Methods

CHAPTER 5

Mapping Interests and Building Psychological Flexibility with the ACT Matrix

In chapter 4 we reviewed the psychological science of behavior change underpinning the Prosocial process. In this chapter, we'll introduce the matrix, a simple tool drawn from acceptance and commitment therapy (ACT)—or, as it's known in organizational contexts such as that of the Prosocial process, acceptance and commitment *training*—that allows you to integrate individual and collective interests within and between groups, work out what needs to be done, and become aware of all the often unspoken ways that the group might hijack its noble goals and undermine its chances of achieving them. Creating such a map gives group members more choices about where they can go and more psychological flexibility for getting there. Using the individual ACT matrix in a group can help create greater trust, psychological flexibility, and perspective taking. To illustrate its use, we'll use a case study of its application to resolve a workplace conflict in a pharmacy.

> *Janine should have been happy and excited. The pharmacy she'd started just five years before was hugely successful and had recently expanded from eight to eighteen staff members. But instead of being happy, she was having trouble sleeping at night, and a few days before she had shouted at her staff in the store in front of customers.*
>
> *Janine described herself as an "old school" manager. She worked long hours and expected her senior staff to do the same, if they wanted to get ahead. What she most cared about, after a long career building a series of successful small businesses, was financial security and being able to leave her pharmacy in the hands of her senior staff while she spent more time with her grandchildren.*
>
> *Those dreams of delegating were in tatters because a conflict between two of her most senior and trusted staff members, Kai and Lateefah, had escalated. Kai was a diligent and quiet highly qualified pharmacist who preferred to work on her own. Lateefah was much younger than Kai and had recently completed her internship. Hoping to eventually own and run a pharmacy of her own, Lateefah had a keen interest in management techniques and read leadership books in her spare time.*
>
> *From Janine's perspective, Lateefah was an ambitious manager who had the potential to become her successor, helping her manage the store and realize her dreams. But from Kai's perspective, Lateefah was a micromanager who didn't treat her with the respect she deserved. For example, in an effort to improve the store's efficiency, Lateefah had developed a "roster" system that required the staff to record what they were doing after each five-minute interval—a system Kai hated.*

The situation between Kai and Lateefah had deteriorated to such an extent that Kai had now taken indefinite sick leave claiming the work environment was "toxic." She had even gone so far as to lodge a compensation claim for being bullied. The two employees had long ago stopped communicating with each other and spoke only through Janine.

As a result of their lack of communication, a recent order had been botched and the pharmacy had lost a substantial sum by purchasing incorrect, date-sensitive medications. Customers were complaining about the service; other staff members were upset and had started talking about leaving the pharmacy. The situation was critical. Janine called Kai and Lateefah together, apologized for shouting at them, and told them that they had to sort out their differences or the pharmacy faced a very real financial threat. Kai and Lateefah promised to try to improve, but because they found it hard to talk about the tension between them, nothing was really resolved, and Kai became more determined than ever to pursue her compensation claim. Ultimately, Paul was called in to help resolve the situation.

Imagine that you've been called in to help. What would you do to try to sort through this mess? Paul turned to the Prosocial process, and the matrix in particular. We'll use this story throughout the chapter to illustrate the use of the individual matrix.

The Matrix as a Multilevel Tool: Mapping Individual or Collective Interests

We generally loop through the matrix at least twice in the Prosocial process. First we look at clarifying individual goals, values, and concerns using the matrix, and then we look at integrating those interests to craft a stronger sense of shared identity and purpose at the group level. Individuals or groups can complete the matrix, and the process can be focused on one's own interests or the collective interests of the whole group. The key distinction to remember is the latter one. When discussing what an individual cares about, we call it the "individual matrix"; when discussing what a collective cares about, we call it the "collective matrix." The following table shows the options for working with the matrix.

	Whose interests is the matrix about?	
Who is completing the matrix?	An Individual's Interests (e.g., What matters to me about…? What do I most care about here?)	The Collective's Interests (e.g., What matters to us about…? What is our shared purpose?)
One person	Individual matrix (completed by a single person)	Collective matrix (completed by one person reflecting on the group)
Multiple people at the same time	Individual matrix (completed in a group setting)	Collective matrix (completed by a group working together)

In this chapter we'll show you how to do the individual matrix and invite you to complete it alone. Once you are familiar with its structure in terms of one of your own situations, you'll be in a position to learn how to use the matrix to help groups improve their collaborative abilities.

Dimensions of Experience

The matrix is a process for thinking about two key dimensions of experience: "toward/away," which reflects the way all animals move *toward* aspects of experience they want more of and *away* from experiences they want less of, and "outside/inside," which represents what is actually happening in the physical world—what people can see, hear, taste, touch, and smell—versus the internal experience of our mind, including our thoughts, emotions, imagery, memories, and the like.[1]

Toward/Away

Speaking loosely, moving "toward" is seeking more of something that we like, whereas moving "away" is seeking less of something we dislike. All animals will move toward food, warmth, and other experiences that sustain life and away from experiences of danger and pain that threaten life. Humans are no different in that respect, except that language and cognition make these toward and away processes much more complex. The things we move toward and away from aren't just concrete, physical things such as food or shelter or predators, but things we think about, remember, or imagine—like the prospect that a sharp-tongued boss might rebuke us if we displease her, or our dreams about retirement.[2]

Generally speaking, the more frequently our behavior is guided by moving toward what we want, rather than away from what we don't want, the happier and more fulfilled we'll be. Moving toward what we care about creates energy and vitality, as well as expands the flexibility and range of our behaviors. Moving away, by contrast, tends to diminish vitality and decrease the flexibility and range of our behaviors. So, for example, if your work is something you want to do because it matters to you, you are much more likely to experience well-being than if you work to simply avoid poverty or the disapproval of others. Over time, people who spend more of their life moving toward what matters have broader and more flexible repertoires of behavior, whereas people who spend their life trying to avoid having difficult or painful experiences have smaller and more rigid repertoires. Indeed, B. F. Skinner defined freedom as being under the control of appetitive stimuli as opposed to aversive stimuli—living in a *toward* mode rather than an *away* one.[3]

Try this exercise to get a sense of how these two modes feel.

EXERCISE: Feeling "Toward" and "Away"

We invite you to reflect for a moment on the activity you are doing right now. Would you say that reading this book is primarily about moving toward something you care about and want more of in life, or is it more about avoiding or getting rid of something you want less of?

We imagine you might be reading this book because you want to move toward being more skillful at managing a group you're part of, or because you want to increase prosociality in the world. But it is at least possible that you're reading this book because you want to appear intelligent at parties and to avoid people calling you shallow, or you're reading the book as part of a class and you'd be punished if you didn't read it.

So ask yourself again, Is reading this book primarily a toward or away move for you?

Now consider some other activities in your life. Consider your school or place of employment. Is the work you do there primarily about moving toward what matters to you—what you value—or away from something you fear or would rather experience less of in life, such as stress, poverty, or the judgment of others?

Note that the toward/away distinction is entirely subjective and context dependent. Only you yourself can tell whether any particular behavior is predominantly "toward" or "away" for you. Speaking up in a group might be a toward move for a person who wants to be more assertive, while it might be an away move for a person who fears people will disapprove of them if they stay quiet. Whatever your particular definition of "toward" or "away" in any given moment, vitality is typically associated with acting autonomously in the direction of one's values, rather than acting under the control of social disapproval or other fears and concerns.

When you were considering the questions in the preceding exercise, you may have noticed that your school or work contains both toward and away elements for you. That is normal, and there is no "right" answer; it's entirely possible that a given activity will have both toward and away elements. For example, writing about the Prosocial process is mostly a toward move for Paul. He cares deeply about collaboration and promoting prosocial behavior in the world. But when trying to get more highly cited publications drives his writing, it feels like an away move—specifically away from the disapproval of university promotional committees! When fear rather than what he cares about motivates his work, it loses its vitality. Knowing this, he can try to focus more on the collaboration and the excitement of doing new research rather than the pressure to publish prestigiously, moving as mindfully as possible through the pressures of the university committees when they can't be avoided.

Outside/Inside

Human beings care about, and worry about, a lot more than animals thanks to our capacity to imagine and tell inspiring or terrifying stories about the past, present, and future. As we explored in the last chapter, humans live in two worlds simultaneously: the world of actual physical contacts and causes, and the world of language and the mind. We can transform the meaning of any situation through the ways we relate to our experience. Winning the lottery can become painful when it causes strife with relatives, and having a leg blown off by a landmine or experiencing cancer can become a source of meaning and joy when the processes of survival and recovery awaken one to the beauty of life. This is not hyperbole—these are actual examples.[4]

Another key capability that's associated with well-being is the capacity to distinguish between what is actually happening and the interpretations and meanings we attach to what is happening. The matrix represents this awareness dimension of experience with a vertical line crossing the toward/away dimension, which is a horizontal line (see figure 5.1).

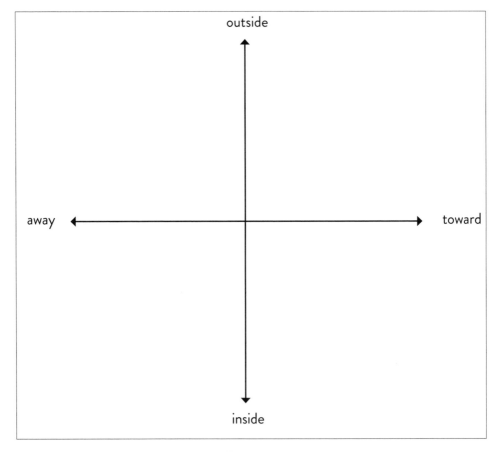

Figure 5.1

As we indicated in our earlier discussions of psychological flexibility, when we learn that what we actually *do* in the world does not have to be a reaction to our internal experience, we can develop greater flexibility and power for responding to situations that are difficult for us. Let's explore what this dimension of experience feels like with an exercise.

EXERCISE: Noticing "Outside" and "Inside"

This exercise is about developing the ability to distinguish actual experience from our interpretation of experience. The very first step is to notice the difference between the two, something we typically don't do. Someone might say something, and instead of really hearing their words all we can think of is what we think they really want or what we will say back to them when we get a chance. Or somebody might do something, and instead of really noticing what they did, we think in terms of our interpretations about why they did it. We even do the same thing to ourselves. If we make a mistake, instead of just noticing the mistake and seeking to fix it, we might slip into old stories about our faults, and where they came from, and beliefs that we'll always make such mistakes—that this is just who we are.

This exercise is meant to help you begin to unpack the differences between actual experiences and the ways your mind responds to them. If you'd like to close your eyes and practice this exercise as a meditation, you can make a recording of this script on your smartphone. Or you can just read it while noticing your experience, if that's easier for you.

> Close your eyes, if you feel comfortable with that, or let your eyes rest softly on a spot in front of you.
>
> First, let yourself notice what it feels like to be in your body, sitting here...
>
> Notice the feel of your bottom in the chair, the pressure of your feet on the floor, the air and clothing against your skin...
>
> And let your attention come now to some particular sensation in your body. It might be a tightness in your shoulders, or a tingling somewhere, or a heaviness, or even a blankness. Wherever it is, it's okay.
>
> Just let your attention settle on one particular part of your body...
>
> Now seeing if you can notice warmth or coolness, heaviness or lightness, density or spaciousness in that sensation...just noticing... Not needing to change anything, or fix anything, just letting yourself be with whatever you experience...
>
> And no doubt you will notice pretty quickly that you have a reaction to this moment. You might notice that you don't like this sensation. It is painful or uncomfortable in some way.
>
> Or you might notice that you *do* like this sensation or that you are thinking that it is great that you noticed it, or whatever...
>
> Or you might just notice yourself remembering other exercises like this, or thinking that you are hopeless at guided imagery.
>
> You might notice that, in addition to your experience in the world, there is the story you build around it, the sense you make of it, whether you like it or want it to stop or whatever.

And now just sitting here for a moment with open awareness. See if, as best you can, you can just stay with an open awareness of whatever your experience is in this moment…

Sometimes people find it helpful to label their experience as it is happening. You might use a label like "outside" to refer to your sensing of the world and "inside" to refer to the thoughts, images, and preferences that you have in response to what you observe. Or you might find it easier to just label these experiences "sensing" and "thinking."

Whatever works for you, taking a moment to sit quietly with your experience, and, as best you can, notice sensing and notice thinking. Also notice the difference between the two—and how easy it is to confuse them.

Good. In a moment I will invite you to open your eyes and reconnect with the room. In that transition, and for as long as you can manage, see if you can maintain some awareness of what you are noticing in your body…just checking in from time to time with what you are sensing. Now opening your eyes and reconnecting with the room…

Once you put the toward/away and outside/inside dimensions together, you get four quadrants:

1. Bottom-right quadrant: Values and other ways of thinking and feeling about what's important to you.

2. Bottom-left quadrant: Thoughts and feelings that get in the way for you that you typically move "away" from.

3. Top-left quadrant: Ways you behave in response to the stuff that gets in the way, whether to avoid it or control its impact.

4. Top-right quadrant: Things you could do (values-based actions you might take) to move toward what is important to you (the values in the bottom-right quadrant).

We'll now explore how you might work with an individual or group of people to have them fill in these four quadrants.

Completing the Matrix

Let's put the toward/away and outside/inside dimensions to work in a practical example that illustrates how you can use the matrix. We'll continue using the pharmacy example to illustrate the process, but we invite you to grab a pen and a piece of paper and follow along using your own example in order to really get what we're talking about. First, draw a cross like the one in figure 5.1 and label its axes, creating the four quadrants of the matrix.

Let's start with the bottom right, what we might call the "values" quadrant. Think of a group that you care about and consider this question: What matters most to you about your group? Write your responses in the bottom-right quadrant. You only need two to three responses for each quadrant, although in our experience you'll be rewarded by taking some time to think about your responses before writing, as initial responses are not always the most important ones. Your values can include relationships, qualities of behavior, or overarching, lifetime goals. What you're striving for here is anything that is important to you. To illustrate what this might look like, we include Kai's responses when Paul asked her to complete this quadrant (see figure 5.2).

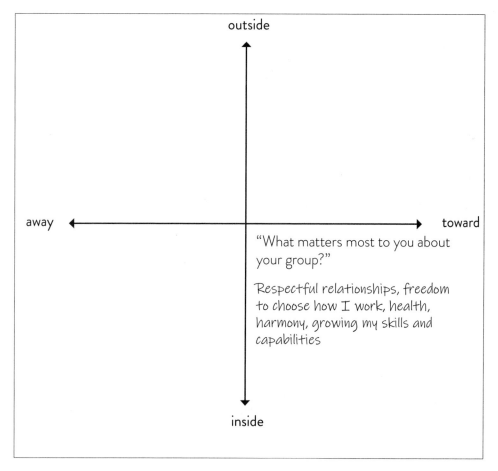

Figure 5.2

One trap to watch out for with this quadrant is responding to social expectations rather than what really matters to you personally. Paul once worked with a nurse who initially wrote "compassion" as a primary value, although he didn't seem particularly enthusiastic about it. When Paul asked him about his response, the nurse realized that compassion was more of a social expectation of nurses than a personal value. When the nurse looked more closely at what created vitality and engagement in his life, he discovered that "adventure" was much more important to him than compassion.

To avoid the problem of social expectations, ask yourself, Does this really matter to me, or is this related to the interests of someone else? This exercise is only between you and the person you can see in the mirror, and the aim is to write down the stuff that really moves you.

There are other ways into this quadrant. You can ask people to reflect on times when they were most alive and felt vital, and then extract key values and overarching goals from their story. Or you can have them identify a guide or hero (see the "Picking a Guide or Hero" exercise in chapter 4). Or you can do some kind of exercise to identify a person's key strengths (such as a strength-finder inventory), as these are often linked closely to values. Or you can do some kind of values card sort or use a list of values.

Next, let's move to the bottom-left quadrant, which is about the inner stuff you would rather not have. You can get at this material with a question like this: What difficult internal experiences show up and get in the way of moving toward what is most important to you? There may be lots of *external* hindrances that get in the way, such as a lack of resources or opposition from others, but we're interested in what shows up *internally* in your mind that hooks you and drags you away from being the person you want to be and acting in the ways that you wish to. Internal hindrances can include emotions, such as fear, or thoughts, such as self-judgments. They can include images or memories of past situations, such as times when you were criticized, or even bodily sensations, such as a trembling stomach.

There is always something to put in this quadrant, because caring about anything creates the possibility of loss, failure, and other forms of pain. Parents fear for the safety of their children, or the ways that they might fail. Authors worry about whether they will have enough knowledge or are entertaining enough to write a successful book. All of us have inner critics that show up and can get in the way whenever we make a significant change in our life. Our mind generates endless material that can hook us and interfere with us moving toward what matters.

Write your responses in the bottom-left quadrant of your matrix. You might find it helpful to distinguish between the types of experiences by writing them down differently, such as putting emotions in all caps (for example, FEAR, ANXIETY, SADNESS) or thoughts in quotation marks (for example, "I might fail," "I'm not good enough"). Again, we provide Kai's responses as an illustration (see figure 5.3).

Mapping Interests and Building Psychological Flexibility with the ACT Matrix 81

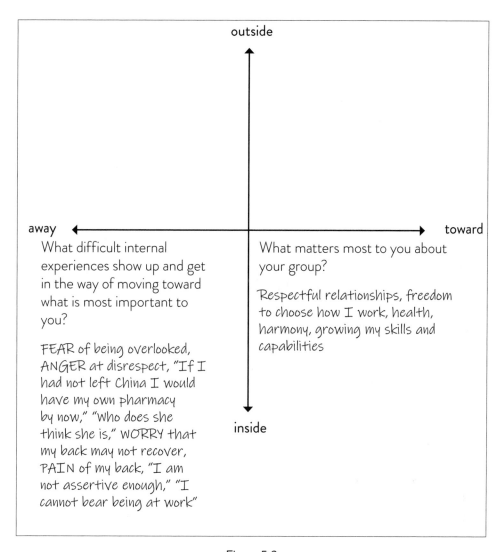

Figure 5.3

Now, moving to the top-left quadrant, answer this question: What can people see you do to avoid or control those difficult experiences? This quadrant is all about what you do overtly to try to minimize the occurrence or impact of the difficult experiences you just mapped in the bottom-left quadrant. We sometimes use the image of a video camera filming our lives to remind people that the answers in this quadrant should be about visible behavior that others can see.

There are three broad classes of behavior that can fit in this quadrant. The most common strategy is to try to *avoid* the painful experience in some way or other. In the presence of fear about a conflict with someone, for example, we might avoid raising the issue. In the presence of fear of failure, we might avoid taking the risk of giving a presentation where others might see how little we actually know. A second broad class of behavior is to attempt to resist, attack, or get rid of difficult inner experiences. In the presence of doubt, for example, we might repeat to ourselves, "Come on, snap out of it," and in the presence of sadness, we might strive to think happy thoughts. Finally, we can seek to ignore or distract ourselves by directing our attention elsewhere. All animals have evolved these fight-or-flight responses for dealing with external threats. But humans amplify these moves as they engage with the internal threats of the mind. Figure 5.4 shows what Kai said she did when she got really hooked by difficult internal experiences.

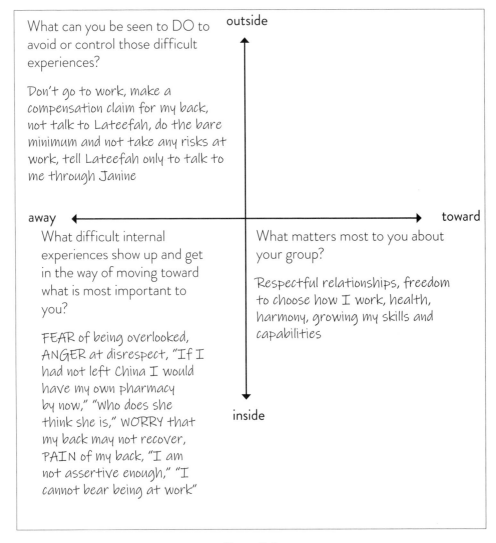

Figure 5.4

For the final quadrant, consider this: What might you do to move toward what is most important to you (even in the presence of difficult experiences)? What are some small or large steps that you might actually take to increase the frequency or presence of what you most care about in your life? Write some down—actually put them on your sheet. Make sure they are things people could see you doing in the world. You have control over how you move your arms and legs, but you don't have control over what others do or your internal experience of the world.

It is at this stage of filling out the matrix, when we start to think of what we might actually do to move toward what we care about, that most of us find our inner critic (bottom left) starts to really heat up. When Kai mentioned that she might take part in Paul's discussion group, she immediately followed it with "But that will just lead to more conflict, and I cannot bear that!" Paul pointed out that this, too, was a thought that hooked her and stopped her from moving toward what mattered to her. He advised her to write it down as a possibility anyway and worry later about whether it would happen and whether she could actually bear it (see figure 5.5).

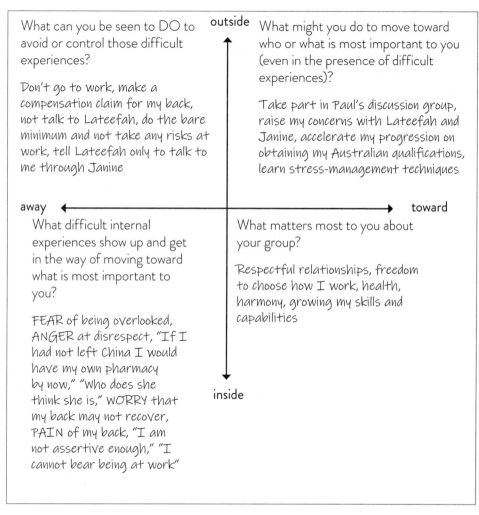

Figure 5.5

Now step back and take a look at *your* whole map. Go ahead and add any additional elements you think might be missing from any of the quadrants. In particular, make sure that you've captured the essence or core of your interests in relation to the group. As you'll see throughout this book, that bottom-right quadrant, with its clear set of values, is a crucial source of energy and motivation for doing the hard stuff that is needed to make a group run better and to enhance prosociality. These values and the overarching goals you derive from them can act as a compass to guide you in your decision making.

This is your personal map of your interests in relation to your group. As we move through the book we'll discuss how you can further explore the collective interests of the group. This map will be an invaluable guide to not losing yourself as you open up to the maps of others in the group because it keeps front and center how you want to be in relation to this group—and the ways you might get sidetracked from moving in that direction. We suggest you keep it available as you read the book.

How Well Is the Top-Left Behavior Working?

It can be helpful at this point to explore how well the various strategies outlined in the top-left quadrant of the matrix are working. You can do this simply by asking three key questions:

1. In the short term, how well is what you noted in the top-left quadrant working to help you control or avoid the difficult experiences in the bottom left?

2. And how well is what you're doing working to help you control or avoid difficult experiences in the long term?

3. And how well is what you're doing working to move you in the direction of what really matters to you?

We've found that it is usually enough just to ask these questions from a place of curiosity, compassion, and open acceptance. Completing the matrix is meant to be a genuinely curious exploration, not a covert way of leading people to preestablished truths. Compassion is important because attempting to protect ourselves from pain is part of the human condition. And open acceptance fits with this process because nothing is good or bad when it comes to what people write in this quadrant—the strategies they use made perfect sense at some point in their life. Many of our strategies for controlling or avoiding difficult experiences, for example, have their genesis in childhood, when we had limited control over our environment and few options but to model our behavior on what we saw others around us do. Unhooking from blame, judgment, or criticism allows us to hold in mind a simple functional question: How well do these strategies work?

Often when people do this exercise, they realize that their self-protective strategies have worked somewhat well in the short term, but in the long term they don't really help them control or avoid difficult experiences, or help them move in the direction of what matters. For example, Kai noticed that not going to work was mildly effective in the short term, helping her to avoid at least some of the difficult thoughts and feelings arising from her conflicts with Lateefah. But when she stayed at home, she found herself ruminating about how things had not turned out as she had hoped and how she had been wronged, or worrying about her future prospects. So she recognized the strategy was not effective in the long term, and it was certainly ineffective in terms of moving her toward what was most important to her, which was thriving as a competent professional pharmacist. Ultimately, staying at home would leave her pretty much stuck where she was. In acceptance and commitment training, this exploration is called *creative hopelessness* because it puts people in contact with the hopelessness of doing more of the same strategies that have not worked, and it also creates an open and creative space for trying out something truly new.

A final question you might consider here is this: Imagine you could choose to relate to your group by either moving mostly away or mostly toward. Which would you choose?[5] This is a powerful existential question: How do I want to live my life? Do I want my tombstone to read "Here lies Paul. He always protected himself and lived a safe life," or "Here lies Paul. He lived large and passionately"? By *asking* people how they want to live life, the essential responsibility and autonomy we all have to choose an answer is retained. Some people will choose self-protection, particularly if they cannot see any other option. But, for most people, this question energizes them with the possibility of trying something different and going somewhere new.

Facilitating the Individual Matrix with a Group

Now we'll discuss how you might facilitate this process with a group, in which multiple people are completing individual matrices at the same time.[6] There are lots of ways to do this. The community of Prosocial facilitators (see https://community.prosocial.world) is constantly discussing different approaches suitable for different contexts. Let's consider a few approaches that can work well.

The Pair-Discussion Approach

One way to facilitate the matrix with a group is to discuss the quadrants in pairs, as detailed in these steps:

1. On a whiteboard, draw the matrix figure, elaborating on the two key dimensions of experience. If you have time, try to do this experientially using the "Feeling 'Toward' and 'Away'" and "Noticing 'Outside' and 'Inside'" exercises from earlier in the chapter or something similar.[7]

2. Beginning with the bottom-right quadrant, the one featuring the question "What matters most to you about this group?" write down your own responses as a model for others to follow. Being willing to be open and somewhat vulnerable will set the tone for the group and maybe give them some ideas about the sorts of responses they might put in this quadrant. If you're not actually a member of the group you're working with, you can simply say out loud some of your own responses to a group you do care about, or examples of some of the things you think members of the group might write.[8]

3. After filling in the first quadrant, invite participants to pair up and share their responses with each other, making sure they know they can choose how much or how little they'd like to reveal to their partner.

4. Continue through the other quadrants, asking the question, illustrating on the whiteboard using your own response, asking the group members to complete their own individual maps, and then inviting them to share what they wrote with their partner.

As facilitator, you need to decide whether the group has sufficient trust and goodwill before inviting members to pair off and share their maps with one another. If there is enough trust in the group, you can even ask for volunteers to share their maps with the rest of the group, writing their responses on the whiteboard as you go.

If you decide to invite participants to share their maps with one another, it's usually helpful to give some brief instructions about how best to listen. We've found the following instructions helpful:

> When your partner tells you about what they have written, your role is simply to listen and reflect what you hear. You can ask clarifying questions, but please don't get into a discussion, make suggestions regarding what they should do, or talk about how what they wrote relates to your experience. The point is to listen to your partner in the way that you would like to be listened to, deeply and fully.

We expand on this form of deep listening in chapter 12, as well as the supplementary chapter "How Can We Make Conversations More Prosocial?" (available for download at http://www.prosocial.world).

This method emphasizes each person's own exploration and perhaps building trust with one or two others. However, from a group development point of view, there is enormous value in being able to create a shared map of the individual interests of everybody in the group.

The Whole-Group Approach

One way to facilitate the whole-group approach is to use a whiteboard to work through the matrix as a group. With each quadrant, the facilitator asks whether anybody would like to volunteer what they wrote down on their private map. In this way, group members choose what is most important, and safe, for them to share with the larger group, and a collective map of individual interests can be created.

The advantage of this approach is that the process can build a stronger sense of shared identity and purpose. Furthermore, group members might recognize differences in their interests and can then start to discuss ways to plan around them. For example, in one group that we worked with, some people valued learning very highly, while others valued producing a very high-quality product. For that group, it was important to resolve the tension between desire to experiment and the need for perfect product quality every time.

Some drawbacks of this approach are that contributions are not anonymous, and the map is inherently incomplete as people only choose to reveal some of their responses. Our experience is that unless there is already a lot of trust in the group, there is the risk that unspoken issues may not surface.

Using Group Software

We've started using digital brainstorming tools to facilitate this stage of the Prosocial process. Such tools can make entries anonymous, an option that can make it more likely that group members will be honest about their perspective, particularly regarding entries on the left-hand side of the matrix. There are tools that automatically collate short entries as word clouds, rapidly identifying shared responses (see figure 5.6); that require individuals to make entries before they can see those of others; that manually group related responses; and that vote on the highest-priority actions in the top-right quadrant (see http://www.prosocial.world/digital-tools). Although we started using these tools while facilitating groups virtually, they have been so useful that increasingly we've been using them face-to-face with groups.

Figure 5.6. An example of a word cloud generated in response to the question "What matters most to you about being part of this group?"

The work in the matrix so far has been about increasing a group's awareness of what is currently going on, how workable that behavior is, and how individuals would like the group to be. We've found that just raising awareness, by having group members complete their individual ACT matrices, is often enough to start producing profound changes in the group as a whole.

But having a map is not the same as taking a journey. You'll also need to lead group members in organized action to create the more cohesive and effective group you want (especially when you move to implementing the core design principles). And usually this involves helping group members practice becoming more psychologically flexible. That is, how can you help everyone in your group develop the skills to move in the direction of the top-right quadrant of their matrix, even in the presence of the difficult experiences they noted in the bottom-left quadrant?

Targeting Psychological Flexibility in the Matrix

The matrix itself fosters psychological flexibility by helping us become more aware of our values and vulnerabilities. It's helpful for members to notice who is making the distinction between toward and away, and outside and inside.[9] You can facilitate this distinction in a group by drawing a circle in the middle of the matrix diagram at the intersection of the two lines and labeling it something like "me" or "noticing." As the group continues to interact, you can occasionally ask them to sort their experience with questions like "Where would you place that on your matrix?" If you do this with a light touch, emphasizing a kind of playful approach in which there are no right answers, this sorting can gently improve understanding and encourage members to implement this way of viewing experience.

There are a huge number of additional techniques available for cultivating psychological flexibility that can readily be integrated with the matrix.[10] In the following sections we'll discuss some of the approaches that we've found helpful in the context of facilitating the Prosocial process in groups, but the widely available ACT literature provides myriad additional methods.

Formal Mindfulness Practices

Short mindfulness practices, such as the "Noticing 'Outside' and 'Inside'" exercise presented earlier, can be useful. They often help provide space for people to nonjudgmentally become aware of their breath or other sensations in their body, to ground themselves in the present moment, before contemplating whether a particular situation or experience is predominantly a toward or away move, for example. When the inner critic is particularly dominant for a person or a group, it is sometimes helpful to focus on exercises that cultivate self-compassion, such as those recorded by Kristen Neff (see https://self-compassion.org/category/exercises).

The Mindful Listening Exercise

Because reflective listening is usually the weakest skill in any group, we often include the following listening exercise at some point in our presentation of the individual matrix.

This very simple exercise allows people to experience the power of mindful listening. You might facilitate this exercise in a number of contexts. If the group is talking about core design principle 3 (fair and inclusive decision making), for example, we may have pairs reflect on times when they didn't have power or weren't included, or on times they stood up for what they believed in. Regardless of when you facilitate the exercise, or the content of the conversation you task people with conducting, the process is the same:

1. Invite people to get into pairs and to nominate a person to be the first speaker (A) and first listener (B).

2. A speaks for five minutes, and B simply listens without saying anything. It's okay if B smiles or nods or responds naturally, but they should not ask questions or take notes. Their role is to listen mainly for feelings, values, and needs—in other words, interests. If A runs out of things to say, they can simply lapse into silence until they think of something else they would like to say.

3. Then for two minutes B reflects back what they heard in their own words while A simply listens. This reflection is all about A's experience. It might involve how B went "beyond the data" to make inferences about what A was thinking or feeling on the basis of their expressions or tone of voice, but it's not about B's reaction to what they heard.

4. A and B then debrief for one minute. They then switch roles and repeat the exercise.

At the end of the exercise, debrief what happened as a group. Usually we begin by asking what it was like to be listened to in this way. We have used this simple exercise in hundreds of workshops, and almost always at least one person in the group reports that they had *never* been listened to in this way, and that it was an extraordinarily powerful experience for them. People often also report that the experience helped them get clearer on their own thinking, and perhaps more committed to what matters to them. In short, speaking to someone who is clearly listening to you is a powerful means for clarifying and expressing your interests.

On the listener side, people usually report that, far from being passive, listening in this way is intense. In most daily conversations, the majority of our attention is on what we plan to say next, rather than on listening to what the other is actually saying. If one truly listens, sometimes they might feel overwhelmed by the complexity of what people say. But giving people permission to not try to remember every detail of what's said, but to focus on feelings, values, and needs, can help cut through this complexity.

There are many useful variants one can employ in this approach. For example, you can use groups of three or four people instead of pairs, and have multiple listeners pay attention to different aspects of what the person is saying (for example, what the person is noticing, feeling, valuing, and requesting). Multiple people sometimes hear different aspects of what is said, powerfully highlighting the reality that our perspectives of the world are only partial.

We like to weave this exercise into the very earliest phases of the Prosocial process, including it while exploring the values segment of the matrix. We sometimes invite people to "Think of an experience or a moment in your life that shaped who you are. You might have been deeply affected by another person or by an experience of some sort. Write down a few key words, reflecting what it was about that situation that so deeply affected you." You can include this prompt as part of a more extended eyes-closed visualization and mindfulness exercise, or it can simply preface a short period of solo reflection. If the group is more visual, you can invite them to use symbols to draw an image of the dynamics of the situation they'd like to address. Then we do the mindful listening exercise with person A speaking about what they have noted or drawn. This process is usually very powerful, often evoking strong positive emotions, sometimes crying, and always a strong sense of purpose and meaning in the room. We find that by structuring the situation to richly reward openness and vulnerability, we also often build trust among participants, the first step toward having a more prosocial group.

Focusing on the "Internal Conversation"

When talking about psychological flexibility in groups that are not psychologically minded, we've found it useful to first present some basic principles of effective communication, such as those outlined in chapter 12 (listening with openness and curiosity, focusing on what is actually happening rather than on evaluations and interpretations, bringing attention back to purpose), and then pointing out that all the same principles can be applied internally in the way that we speak to ourselves. Consider Kai's harsh internal dialogue: "I am not assertive enough" or "I should never have left China." She is more likely to act flexibly and move in the direction of her values if she can greet those inner "voices" with openness and curiosity and be self-compassionate toward them. After all, the voices only function to try to protect her from harm or to improve her relationships. If she greets them with further self-condemnation, she will simply extend the cycle of self-criticism and fear.

Other Tips and Tricks for Facilitating the Individual Matrix in a Group

You have lots of choices for facilitating the Prosocial process with your own group. In this section we address questions that facilitators new to this process often raise.

How do I explain why we're doing the matrix? The reasons why a group may be interested in doing the Prosocial process vary widely, but they are likely all variants of a broad desire to improve cooperation and group effectiveness. To create truly high-performing teams, it is essential to know what individuals care about. Thinking back to multilevel selection theory (chapter 1), the individual matrix creates a map of both the pull toward cooperation (generally the collaborative elements on the right-hand side of the matrix) and also the tendency to act out of self-interest (some of the self-protective elements of the left-hand side). The individual matrix both increases self-awareness of personal interests and provides a safe way to talk about the usually undiscussable self-protective thoughts, feelings, and behaviors that can get in the way of collaboration. But the matrix goes well beyond self-awareness. When people share their individual ACT matrices in a group (as we described previously), you can use those matrices to begin the crucial work of taking the perspective of others, increasing trust, and knitting together individual and collective interests.

Can I use the matrix when group members don't really trust one another? Sometimes facilitators are concerned that participants won't share personal material because there's not sufficient trust within the group. If this is true of your situation, it may be a sign that your group is not quite ready to use the Prosocial process yet; you'll need to work to secure their buy-in before going through the matrix and the larger Prosocial program, making sure that everyone in the group sees the need for change and is on board with the Prosocial process. Here are some issues to consider.

- *Get really clear with the group on why doing the Prosocial process matters:* If the group has a strong enough reason to do the Prosocial process, then members will be more willing to take risks. But if they see the process as just another "team-development exercise" without any clear reason for doing it, the process is unlikely to succeed.

- *Focus on creating an environment where vulnerability is met with positive reinforcement:* To trust another person, one must have the expectation that their actions will be met with a positive response, or at least not a highly punishing response. The general behavioral principle guiding our work in groups is to evoke openness and then make sure that openness and vulnerability are richly reinforced both by the facilitator and by other members of the group.

- *Talk about what is needed to create trust in the group:* We often talk about the importance of trust at the beginning of the process. Sometimes you can start this discussion with simple questions, such as "What should we do as a group to have quality conversations?" and "What should we do as a group to make sure that everyone feels safe to speak up and be honest?" The answers might include establishing guidelines for how you want your group members to respond to one another (for example, don't criticize or judge, just listen or acknowledge) and limiting any responding that diminishes trust. Then, if anyone violates these cocreated guidelines, you can simply remind the group of what they said and ask whether a particular behavior is in line with those guidelines.

- *Model vulnerability:* Often just being willing to share, as the facilitator, some of your own hooks in relation to the group or the facilitation process is enough to model the kind of vulnerability you're looking for. We almost always use real and current examples of our own to help populate the matrix, either on a whiteboard or online.

- *Have conversations directly with people who are in power in the group:* Members of the group who have more power than others can dramatically influence whether trust develops. Do whatever you need to do to ensure that they participate in ways that reinforce rather than punish openness and vulnerability.

- *Consider using variants of the matrix that are less threatening:* Our colleague Magnus Johansson developed a variant of the matrix that uses the labels "more of" and "less of" rather than "toward" and "away." He finds that these terms are less threatening to some groups.

What question should I ask for the bottom-right quadrant? The question you ask for the bottom right determines the flow of the rest of the matrix. We suggested asking "What matters most to you about your group?" but you can focus this question any way that seems relevant. You might, for example, just ask this generic question: What is most important to you in your life? This question invites a broader integration of all of the interests of group members compared to our recommended question, which focuses on group-relevant issues and values, and might yield the most useful information for enhancing group collaboration.

How do I maintain curiosity, compassion, and openness as the facilitator? The key to successfully facilitating the matrix is to remain curious, compassionate, and open to the perspectives of everyone in the group. This is their map of their world. Ultimately it doesn't matter whether they use words that seem odd, or put things in the "wrong" quadrants, or respond with an attitude when you ask for a behavior. Those phrases and placements and behaviors make sense to them. This exercise is all about increasing the group's awareness of the processes of variation, selection, and function of behaviors. It's all about the function of the behaviors that the group and its members are engaging in, and the workability of those behaviors, not "the truth."

To orient yourself to this stance, it's helpful to focus more on the *function* of what is said rather than its specific form. For example, if someone insists that keeping quiet and avoiding conflict are important and useful parts of their approach to managing difficult thoughts and feelings, or issuing harsh coercive commands is an effective way to motivate people, it's usually best to roll with it. More than likely the sense of safety your acceptance creates will allow them to reexamine their assumptions and beliefs throughout the process.

If you're collecting responses on the whiteboard, it's usually best to use the actual words that are proposed (especially at the beginning of the exercise), or at least to ask permission before using an alternative, to ensure that you're not putting words in anyone's mouth. Sometimes group members ramble on and think aloud about their response. In such a situation, you will have to paraphrase. It might be helpful to say something like this: "Let me see if I have this right. It sounds as if you are saying that X [for example, honesty, productivity, and so on] is very important to you?" Again, to ensure you're capturing and not changing what they mean, usually the best thing you can do is simply reflect back as best you can what you think the person is saying or feeling.

What if someone gives me an answer that doesn't fit in the quadrant I'm working on? Again, don't get too hung up on what goes in which quadrant. The main function of this activity is to get the issues and perspectives out of your group members' heads and onto paper. Worrying about whether something should be toward or away, or both, or whether something belongs in the inside or outside world, will not help, and doing so may make the exercise itself aversive. The most helpful stance you can take, in relation to both your own and to others' matrices, is open, curious, receptive awareness.

Imagine, for example, that you ask someone what they could be seen to be doing to move away from pain, and they say something like "acting defensively." This isn't really an observable behavior—but you can try guiding them to clarify what behaviors "acting defensively" entails. You might reasonably ask, "And what would we actually *see* when you are acting defensively? What might that look like if I was filming you with a video camera?" Hopefully this will do the trick to elicit a more specific, behavioral response. If it doesn't, however, the best thing to do is to simply accept their response and move on. Ultimately, it is much more important to build warmth, acceptance, and trust in the room than it is to educate someone about the tool.

How do I manage strong emotions in the room? Strong emotions can arise at any stage of this process, and they are particularly common when group members explore the workability of their past responses to difficult experiences. How you handle these emotions will depend on context, but in general it's helpful to remember the following:

- When the situation allows, it's often helpful to state back to the group member what emotion you think they might be experiencing (for example, "It sounds like that is very upsetting for you"). This lets people know that it's safe to have strong emotions, and that they can safely express them without judgment or censorship.

- Maintain a stance of open, nonjudgmental awareness.

- Give people permission to share as much or as little they want at every stage.

- Notice your own tendencies to avoid or shut down conversations that include emotional talk. If you tend to avoid this kind of conversation, practice having more emotionally oriented conversations in your personal life, and notice yourself developing your own capacity to express and listen to emotions.

- Remember that filling in the matrix is all about mapping the experiences of group members. Experience is what it is about. You don't need to buy into the content in order to express empathy. If someone says, "I am furious that John asked us to do this project," you can just note "fury" on the map. And if someone reports that they often have the thought "I am incompetent," you don't need to buy into the content by, for example, reassuring them that you believe they are competent. Instead, for the purposes of this mapping process, it's more helpful to reflect what was said; empathize with the feeling, saying something like, "I can see it is really difficult for you to have that thought"; write down "thoughts of incompetence"; and move on. The point in this example is that you reflect how the thought functions for the group member, whether it helps or hinders their pursuit of their interests, and not challenge whether or not it is valid.

- Most of all, maintain an open attitude of dispassionate curiosity. That will help establish a sense of safety that in itself is helpful to most groups.

So, What Happened at the Pharmacy?

You might be wondering what happened at the pharmacy we mentioned at the beginning of this chapter. The conflict in this situation was getting in the way of even basic functioning of the group. Paul decided it wouldn't be effective to start with a group session, so he first worked through the matrix with Janine, Kai, and Lateefah individually so that they each had a map of their own perspectives, values, and concerns. He then convened a session in which they used those maps as the basis of a process to work through the immediate conflict. That session was a huge success, as each person was able to see the priorities and concerns of the others. As they left the building, Paul saw Janine, Kai, and Lateefah hugging one another, ready to carry on with the rest of the Prosocial process in order to develop sustainable long-term collaboration.

Conclusion

Completing the individual matrix is the first step in the Prosocial process because it helps individuals clarify and articulate their personal interests. The matrix also helps bring to the surface the inevitable internal concerns and barriers that often go undiscussed and interfere with group members behaving the way they want to.

As group members share the results of the individual matrix, it's common for a sense of connection and support to emerge naturally as people realize that they are not the only ones in the group who hold particular values and vulnerabilities. It is an added benefit that the collective matrix that will follow is functionally identical to the individual matrix, framed from the perspective of the collective rather than the individual. Thus as the members create a map of individual interests, they are preparing to take steps as a group that will honor members' individual goals and values, as well as the struggles they all experience in pursuing those interests. This is key because it allows for authentic discussions about how they can integrate individual interests with those of the group, which will be important when they use the matrix to help build shared identity and purpose (core design principle 1) and explore each of the other core design principles in more depth. In the next chapter, we describe how the various modules of the Prosocial process can be combined to respond to the needs of different groups.

CHAPTER 6

Modules and Pathways for the Prosocial Process

So far in this book we've explored the core design principles (CDPs) as a multilevel framework for improving cooperation within and between groups, and we introduced the matrix as the main tool for integrating individual and collective interests to help implement the CDPs. We are now well on the way to having a practical method for improving collaboration within and between the groups you care about. We designed the Prosocial process to be modular. It contains six interlocking modules, which groups can use separately or in combination:

1. Measurements for assessment and diagnosis
2. The individual matrix
3. The core design principles
4. The collective matrix
5. Goal setting
6. Measurements for evaluating change

We have already talked at length about the individual matrix (chapter 5) and the core design principles (chapter 3). In this chapter, we'll briefly describe the remaining modules of the Prosocial process: tools for assessing what's going on (1), the collective version of the matrix (4), goal setting (5), and processes for evaluating change (6). We'll elaborate upon each of these modules in the remainder of the book. This chapter will also discuss the different ways you might arrange the modules to create an effective intervention with a group or group of groups. Our aim in this chapter is to give you an overview of options for the whole process before proceeding.

Measurements for assessment and diagnosis: It's important to know what's going on with the group before you begin your intervention. The Prosocial process includes brief and comprehensive tools to assess both how well individuals are doing and the collective functioning of your group. Measurement is important for creating a shared perception of the current state of the team, for incorporating multiple perspectives, and for comparing your group with other relevant groups. Although optional, we highly encourage users of the Prosocial process to take advantage of the range of measurement options available at http://www.prosocial.world; they'll support the work of you and your group.

The collective matrix: Once people have learned how to use the individual matrix, you can reuse the same tool to consider collective interests by simply changing the question in the bottom right, from somethings such as "What matters most to me?" to "What matters most to us?" If the individual matrix helps individual interests, concerns, assumptions, and goals to surface, the collective matrix helps knit those together as a shared identity and purpose (core design principle 1; see chapter 7). Furthermore, the group can use the collective matrix to explore how and why each of the CDPs matters, as well as the psychological and behavioral barriers that arise when the group considers options for implementing the principles (see chapters 7 through 12).

By developing skills in "matrixing" at both the individual and collective levels, group members are embodying the activity of engaging and integrating individual purposes and interests with collective purposes and interests. In many if not most groups, this sort of integration is disrupted because people are simply expected to toe the line for the good of the group. In some groups, such as community groups, for example, the collective interest might become almost invisible, and people's individual preferences tend to dominate. Conducting the individual and collective matrices allows groups to integrate individual and collective purposes and interests.

Goal setting: While learning about the principles is useful, true transformation will only occur if groups actually change what they're doing. Your work with the collective matrix to build shared purpose and explore each of the principles will produce a large list of potential goals for committed action for the group, which are anchored in organizational and individual values. Toward the end of the Prosocial process, it's important to prioritize these goals to turn them into committed actions. Usually this involves consolidating, prioritizing, and delegating actions that the group has both the capability and the motivation to execute. We will cover goal setting in chapter 15.

Measurements for evaluating change: The Prosocial process is itself an evolutionary process. Using the relevant measurement tools available at our website (http://www.prosocial.world) to assess the effects of change will contribute to the collective effort of continuously improving and validating the Prosocial method. We have a range of survey tools, from those that take a few minutes to complete to comprehensive assessments of individual and group functioning. Facilitators find that these tools add tremendous value to their interventions. The website will eventually be designed to encourage global research on the Prosocial process in different circumstances using both qualitative and quantitative approaches. We also expect it will feature a range of public domain and custom-made measures appropriate for tracking individual and collective changes in such aspects as effectiveness, values and goals, well-being, flexibility, trust, commitment, and satisfaction.

Pathways

Every intervention will be different depending upon the circumstances. Here we describe three typical ways we have combined the modules described above. Any of these approaches, which we call pathways, can work well in the appropriate context.

Pathway 1: The "standard" pathway. This approach is suitable for most groups, including those with limited time. It's also the approach that we use on our website (http://www.prosocial.world), so you can implement the approach using the website as a guide. While we have seen this approach have some benefits when implemented as three ninety-minute group sessions, having three half-day workshops (or one full day followed by a half day some time later) is preferable. The general approach consists of the following steps:

Before First Session

1. Assess individual and group aspects using the Prosocial survey (http://www.prosocial.world) to evaluate and inform change.

At First Session

2. Implement the individual matrix in a brief and general form with the emphasis on the group simply getting used to the structure. This run-through doesn't need to focus on the group and could simply start with an invitation like "Jot down two or three things that matter most to you in the bottom right."[1]

3. Spend most of the session doing the collective matrix to build an initial awareness of group purpose and identity, as well as other aspects of group functioning.

Between First and Second Sessions

4. Collect perceptions and stories about the CDPs using the survey available at http://www.prosocial.world.

At Second Session

5. Discuss the CDPs survey results.

6. Review how implementation of the CDPs might be improved, and any barriers to improvement.

7. Set short- to medium-term goals.

8. Prioritize and plan actions.

Between Second and Final Sessions

9. Implement planned changes.

At Final Session

10. Review progress (including reminding the group of the shared identity and purpose emerging from the collective matrix). Possibly use the collective matrix to have the group reflect on the reasons for its rate of progress implementing the CDPs.

11. Problem solve how to improve the application of the CDPs.

12. Commit to implementing them, and plan for continuous improvement.

After Final Session and Going Forward

13. Measure individual and group aspects to evaluate change.

Pathway 2: The "interpersonal" pathway. This approach is best for groups that want to start by building stronger interpersonal relations, particularly when past efforts to improve relationships were superficial or ineffective. While this approach can be adapted as needed, we've found that at least three half-day sessions are usually needed to produce transformational change in groups.

Before First Session

1. Assess individual and group aspects to evaluate and inform change using the surveys available at http://www.prosocial.world.

At First Session

2. Implement the individual matrix in more extended form, perhaps using a tool that's recommended at http://www.prosocial.world/digital-tools to fully and anonymously articulate individual interests in relation to the group.

3. Explain the idea of the collective matrix as a tool for elaborating core design principle 1, shared identity and purpose, and how to work on it during the period between the first and second sessions (see next point).

Between First and Second Sessions

4. Encourage participants to complete their own responses to the collective matrix for core design principle 1 using the template available at http://www.prosocial.world, or a digital brainstorming tool such as those recommended at http://www.prosocial.world/digital-tools.

5. Collect perceptions and stories about the CDPs (use the survey available at http://www.prosocial.world).

At Second Session

6. If you used a template, use a whiteboard to complete the collective matrix. If you used a digital brainstorming tool, review the output and add any additional points that arise.

7. Review the collective matrix. Aggregate the responses and vote as needed to prioritize the values and goals found in the right-hand side of the matrix.

8. Review how well the CDPs have been implemented in the group. If there's time, use the collective matrix to clarify the principles and determine potential barriers to their effective implementation (see chapters 7 through 12 for examples of when the collective matrix can be used to better understand and implement the principles).

9. Set short- to medium-term goals.

10. Prioritize and plan actions.

Between Second and Final Sessions

11. Implement planned changes.

At Final Session

12. Review progress (including reminding the group of the shared identity and purpose emerging from the collective matrix). Use the collective matrix where necessary to have the group reflect on the reasons for its rate of progress implementing the CDPs.

13. Problem solve how to improve the application of the CDPs.

14. Commit to implementing them, and plan for continuous improvement.

After Final Session and Going Forward

15. Measure individual and group aspects to evaluate change using the tools available at http://www.prosocial.world.

Pathway 3: The "strategic planning" pathway. Some groups are less psychologically minded (for example, as is sometimes the case with the military or technical disciplines, such as engineering or project planning). With groups like this, we've found it helpful to begin by discussing the CDPs that are seen as less personal and more task focused. The collective matrix can then be introduced to assist with core design principle 1 (shared identity and purpose). If this exploration goes well, people often quickly realize that their personal interests and patterns of responding need to be mapped and understood; in that case, there is a mandate for doing the individual matrix. This format requires a minimum of two half-day sessions.

Before First Session

1. Assess individual and group aspects to evaluate and inform change.

2. Collect perceptions and stories about the CDPs (use the CDPs survey available at http://www.prosocial.world).

At First Session

3. Review and collectively make sense of the results of the CDPs survey.

4. Implement the collective matrix to build an initial awareness of purpose and identity (core design principle 1) and/or other aspects of group functioning (core design principles 2 through 7).

5. Discuss and plan actions (top-right quadrant of the collective matrix).

Between First and Second Sessions

6. Implement actions and plan for continuous improvement and review.

   (*Optional second session:* Implement the individual matrix in brief or extended form only as needed, such as when a group encounters difficulties implementing the principles.)

At Final Session

7. Review and further plan actions, perhaps using the collective matrix to examine progress implementing the principles.

After Final Session and Going Forward

8. Measure individual and group aspects to evaluate change using the tools available at http://www.prosocial.world.

In situations where it seems that the individual matrix would generate more resistance than useful information, it might be appropriate to skip it altogether. However, we see that as a substantial loss. One group we worked with had a leader who did not appreciate the value of what he saw as "soft skills," and so he refused to do the individual matrix with his group. Although using the collective matrix alone worked up to a point, the Prosocial process never quite took hold. In the end the group limped along relatively ineffectively because organizational inflexibility flowed from the failure to work on psychological flexibility. From our perspective, this was a situation in which the defensive behaviors of the individual manager actually interfered with the effectiveness of the group.

Of course, these pathways are merely guides, and you will need to decide how you want to implement the Prosocial process given your situation. Different circumstances may require additional modules and processes. For example, some groups may need to engage in a scenario-planning exercise to envision a range of possible futures that might affect the group.[2] We describe an example of such an exercise in chapter 15 as a prelude to goal setting. Even though the individual matrix is an excellent icebreaker in its own right, sometimes groups need to first engage in other trust-building exercises, such as shared projects and social events, before they can even begin to work with the individual matrix.

Now that you're oriented to the Prosocial process, we're ready to begin with the first, and most important, core design principle: shared identity and purpose.

CHAPTER 7

Core Design Principle 1
Shared Identity and Purpose

In the introduction, we mentioned how the Prosocial process was used to help slow the spread of Ebola in Sierra Leone. Hannah Bockarie, an ACT-trained social worker working in the Bo District, used the Prosocial process during the height of the Ebola outbreak to come up with a ritual that allowed people to honor their traditions of washing and praying over the dead while still avoiding infection. The people themselves called it a "reparation ceremony"—an act of repair directed toward their ancestors and culture.

What happened in the Bo District was only possible because the members of the community felt a sense of belonging, and almost all of them identified with the purposes of both honoring their dead and preserving the lives of those in their communities. When we ask people about the most effective, cooperative groups of which they have been a part, they almost always say that those groups had a strong sense of shared purpose and created a sense of belonging and shared identity.

Ostrom's earlier work focused on groups managing common-pool resources. These groups had a clear purpose built in to their work: to manage a resource. Therefore, she originally framed the first core design principle as "clearly defined boundaries." When your shared purpose is sustainably managing the commons, deciding who has the rights to share the resource and make decisions about it defines the group. But while being clear about group membership and entitlements is critical for all groups, Ostrom later recognized—in her work with David—that the essence of this principle can be framed in a more general and flexible way in terms of shared identity and purpose.

In this chapter, we explore why shared identity and purpose are important and how you can enhance these aspects of groups. We begin by focusing on the more general idea of shared identity, then dive into the more specific notion of shared purpose. As you'll see, shared identity and shared purpose are closely related ideas, which is why we feel okay combining them as a single principle.

Shared Identity and Group Cohesion

Have you ever been part of a group in which you felt you truly belonged? A group with members you trusted and felt close to? A group with a purpose that you were so committed to that you had no thought of leaving? Paul's most recent experience of this state involved a group of three families and three guides rafting the Franklin River in the wilds of Tasmania. Every day the purpose was crystal clear: get further down the river safely while having fun. And every day the group worked perfectly, mixing who went in each of the three rafts and cooperating through the challenges of working out how to negotiate each new rapid, eddy, or cascade.

Such experiences, while rare, often stand out as peak experiences for the thousands of people whom we've asked these questions. Being part of a group with strong bonds of commitment, trust, and shared purpose is deeply, deeply satisfying for human beings, particularly in the age of *Homo economicus*.

It's hard to define such experiences with a single name. You might wish to talk about a sense of shared belonging, mutual trust, or attraction and commitment to the group. Much of the research in this area describes shared identity as "group cohesion" (literally the degree to which the group sticks together and is unified in the pursuit of its goals) or, at a broader level, "social capital," describing the quality of networks of relationships across an entire society.

We have chosen to use the term "shared identity" in the label for this core design principle because it captures a critical aspect of the Prosocial process—when members come to proudly see the group as related to "who they are." In using the individual matrix and then the collective matrix, discussed later in this chapter, the Prosocial process is designed to integrate individual and collective interests so that each member is more likely to identify with the group: they are happy to say that the group represents something they care about, they believe in the group's goals, and they are committed to staying with the group. Behaviorally speaking, we're trying to create a shift from "What's in it for me?" to "What's in it for *us*?" in which the "us" includes me as an important part but is not limited to me.

We'll have much more to say in this chapter about how this can be achieved. For now, we want to briefly turn our attention to considering why this topic is so important for small groups. Because we'll be citing research that's focused on group cohesion in this section, we'll use that term here.

There are two faces to group cohesion: social cohesion and task cohesion. The former is all about the quality of relationships in the group. Do people like others in the group? Do they trust them and feel a sense of affinity with them? The latter is all about the quality of commitment to the shared purpose. Both are necessary for high-performing groups. You might really believe in the goals of the group, but if you don't get along with others in the group, your personal performance and the performance of the group will be substandard. Getting along well with others in the group not only creates a sense of belonging, it increases commitment to stay with the group long term. Conversely, you might get along famously with your group but not agree with or be committed to what it's doing. Once again, your group won't perform as well.

Individuals are happier when their groups are cohesive. Group cohesion also protects group members from stress and depression. Group members even feel more personal meaning in life when there is a strong sense of belonging in a group, because it expands their awareness to include a cause larger than themselves.[1]

Group cohesion also has positive effects at the collective level. More cohesive groups perform better over time and, interestingly, the opposite is also true—better performance also enhances cohesiveness.[2] So it is possible to get a virtuous cycle in which improvements in group cohesion enhance performance, which in turn further improves cohesiveness. Of course, this cycle can also work the other way!

Members of more cohesive groups are more likely to trust one another, share important information, and coordinate their actions more effectively. And members of cohesive groups are more likely to be motivated and therefore more likely to work hard toward group goals, as well as be more loyal and more likely to stay in the group.[3] Interestingly, all these positive effects of group cohesion are even more important for self-managing groups (see chapter 13 on core design principle 7, authority to self-govern) for which having a sense of cohesion has been shown to be a more important motivator than money.[4] While the core design principles are important on their own, we think it's their combination that is really powerful.

How to Build Shared Identity and Group Cohesion

Shared identity and group cohesion are not static traits of a group. They are constantly in flux. A group might be proceeding nicely and then someone violates the trust of the group and cohesion disintegrates. Or the group might attend a retreat and cohesion increases substantially. In general, like trust, cohesion is slow to build and easy to fracture. But there are definitely things you can do. Here are a few ideas that might be helpful for your context.

Use the individual matrix before the rest of the Prosocial process. While groups can evolve to be deeply prosocial and altruistic, people will only stay in groups when belonging is reinforced. You can easily check whether members feel like they belong by asking them to consider and discuss this question: What are the benefits of belonging to this group for me personally? This need to belong is why we often start our process with the individual matrix (see "Pathway 2: The 'interpersonal' pathway," in chapter 6). You want to make sure that individual needs, values, and fears are included in the conversation. It's essential to build this foundation of trust and to take care of individual needs before starting the vulnerable work of discussing shared purpose. And a group without authentic expression will not engage people deeply, leading members to discover superficial and inauthentic purposes. When we care about the same things, it's easier to have a strong sense of shared identity.

Utilize the power of small groups. Small groups are usually more cohesive than larger groups. Larger groups are more prone to social loafing if members think that their loafing will not be noticed or will be tolerated (see core design principles 4 and 5, monitoring agreed behaviors and graduated responding to helpful and unhelpful behavior, respectively). When behavior is monitored (core design principle 4), which is easier to do in small groups, social loafing is less likely to occur. If you have a very large group, look at ways you might divide it into smaller connected groups that can build more social cohesion.

Clarify membership. It is important to be clear on who is in and out of a group. Steve once used the Prosocial process with a group creating alcohol-free social events for college students. When he worked on the principles with the group, issues with core design principle 1 stood out: a fair number of students liked to attend the events, and did so regularly, but never thought of themselves as part of the group. The vague boundaries of membership were causing frustration and undermined both the cohesion and the effectiveness of the group. All of the other design principles suffer when the first principle is weak. For example, those who were actually organizing the events began to feel used: there was not an equitable distribution of contributions and benefits (core design principle 2; see the next chapter). Only when the group clarified that repeated attendance at the social events required group membership did the project begin to prosper due to increased group participation.

Shared identity can be difficult to attain in modern dynamic organizations. People may be included just for parts of a project, or to provide consultation, but not be central to the team. Sometimes groups are nested within other groups, and sometimes membership changes rapidly. Consider Enspiral, a network of social entrepreneurs working together to support one another in building socially conscious business initiatives. Early in the initiative, it became clear that there needed to be two categories of belonging: *members*, who would take responsibility for managing the network and determining its strategic direction, and *contributors*, who would form the network itself, connecting with one another as they needed to in the course of working on the business initiatives they were spearheading (but not necessarily involving themselves in the day-to-day management of the foundation). Whereas members formed an inner circle with correspondingly larger rights and responsibilities, contributors formed an outer circle with fewer rights and responsibilities.[5] To join the network, contributors only had to be invited by one member; new members, however, had to be approved by all the current ones. Such clarity regarding the nature of membership both builds a sense of belonging among members and contributors and enables all the other activities of the network to take place more harmoniously.

Make space for the personal. Group practices that give people space to share about their lives outside the group are usually helpful for building group cohesion. Team-building activities that include nonwork-related activities can give members a chance to shine and to share aspects of themselves they might not normally share. Group cohesion grows with trust, and trust grows with seeing ones' fellow group members as whole people, not just robots doing a task. In the Prosocial design team, we've found that the simple process of doing a "round" to check in with every person in the group at the beginning of a meeting not only helps people concentrate on what needs to be done, but it usually provides a context for better understanding what others in the group are experiencing and how we all sometimes face similar struggles. Of course, the matrix can become a shared language in a group for talking productively about some of these struggles.

Improve communication skills. Developing good-quality listening, questioning, and goal-setting skills (discussed in chapter 12 in particular), as well as providing good information, all help create cohesion by improving channels of communication.

Share reflections. After a project has been completed, coming together as a team to explore what worked well and what can be improved can prove extremely useful. Doing so not only helps with learning, it also enables members to share information, revisit the group's purpose, and build a shared sense of progress and identity, particularly when completions can be accompanied with a celebration!

Use "onboarding" processes. Making sure that new members are welcomed into the group and given all the information they need can increase cohesive relationships and also clarify shared purpose. Even small acts to create a sense of belonging among newcomers can have profound long-term effects upon cohesiveness. One study invited minority college freshmen to read and internalize encouraging messages from more senior students telling them about how the first year can be tough, but that it's temporary. This tiny intervention improved academic performance, health, and well-being for years afterward.[6]

Create similarity in the context of diversity. Groups tend to be more cohesive when their members are more similar. Sometimes groups work to increase perceived similarity by adopting shared uniforms, rituals, or ways of talking about events, such as shared stories or jargon.

But just focusing on increasing similarity is risky. First, increasing member similarity through selective hiring or induction processes can increase conformity rather than creativity and healthy diversity. The focus of the Prosocial process is to create groups in which individual identities are strong and vibrant but well aligned within the context of the group. We are wary of uniforms, team-building exercises, rituals, or the like being introduced to increase cohesion through diminishing individuality, unless these changes are initiated by members.

You can minimize this risk of conformity by finding some way to acknowledge and celebrate diversity within the context of the overall shared purpose of the group. You can celebrate the enhanced creativity and problem solving resulting from diversity, or the unique strengths that each person brings to the group's shared purpose.

Extremely strong within-group cohesion can be problematic for another reason; it can work against good relations with other groups, a violation of core design principle 8, collaborative relations with other groups. Because of the two-way street of symbolic relations (see chapter 4), constructing a sense of "we" invariably attaches a sense of "them" to those who are not part of the group. The force of inclusion creates the force of exclusion.

For example, suppose a group member is a member of a political party. If this person says, "I am a Democrat," how does that category nest with the term "Republican"? Can both identities nest comfortably underneath overarching categories, such as "citizens" or "intelligent people trying to solve problems," or have they become disconnected from any overarching and unifying identity? When "I am a Democrat" becomes disconnected from "I am a citizen," this group identity may make it more important to "win" at politics than to serve the needs of citizens. Cooperation between groups is reduced. For example, a Republican might believe in affordable health care but be unwilling to support a plan to produce exactly that because her group identity gets in the way.

One approach to lessening this sort of intergroup identity conflict is to work so that "us" and "them" together constitute *another* "us" at a higher level of organization. Studies in which participants have been asked to pay attention to how their group is different from others (for example, "college students" versus "not college students"), or to the superordinate features that unite groups (for example, "all residents of the same city"), have shown that when people take on a superordinate identity, they are less likely to be free riders and more likely to make prosocial contributions.[7] Authorities make use of this strategy when they introduce awards such as "tidy towns" that make salient a new level of organization that may provoke cooperation—in this case, a township identity that unites the whole group.[8]

Create a strong sense of shared purpose. If we want to take advantage of the many benefits of a strong sense of identity, while still celebrating individuality and preserving group-to-group cooperation, it helps to focus on shared *purpose*. Furthermore, finding shared purpose is by far the most powerful way to build shared identity. We left shared purpose to last because we devote the rest of the chapter to how you might approach it.

The Importance of Shared Purpose

There are many practical benefits for groups to have a shared sense of purpose. Let's review some of them.

Shared purpose creates a sense of group identity and belonging. When people share a purpose, they are more likely to care for the group and feel a sense of belonging.[9] This makes them more likely to engage in prosocial behaviors, such as sharing information and helping others. You can see this effect in overdrive when a community experiences a natural disaster, which can galvanize them to help others achieve the clear shared purpose of responding effectively to the crisis.

Shared purpose guides individual actions, empowering individuals at all levels of the group. Every possible action a group might take can be evaluated in light of how it will move the group toward its shared purpose. When everyone is clear on the purpose of the group, it's much easier to predict others' behaviors and align actions in the group.

In some circumstances, a clear sense of shared purpose is so powerful that it can substitute for a formal, hierarchical leader role by guiding behavior in ways that are aligned and coordinated. For example, the world's largest tomato-processing company, Morning Star, uses peer-to-peer contracts (known as "colleague letters of understanding") in which each person outlines how they will contribute to the company's success by achieving their personal goals. Rather than relying on leaders telling people what to do, colleagues use a negotiation process built around a shared purpose to determine their activities. Rich Bartlett, cofounder of Loomio, software that assists groups with the decision-making process, writes, "If you articulate your purpose, that becomes the ultimate authority in your organisation, and the criteria by which you design your structure and processes."[10] Morning Star's approach to managing staff embodies this notion.

Enlightened leaders have long appreciated that understanding and endorsing a shared purpose empowers group members to act without being explicitly told what to do. When purpose becomes the boss, there is a shift from coercive "power-over" relationships to cooperative "power-with" relationships, and the team is pulled, not pushed, toward the future.

Shared purpose creates motivation. When people believe in the vision, mission, and values of the group, they are much more likely to exert discretionary effort to support collective efforts. This is a constructive process, not a passive one. When different possible actions that are available in the present moment are connected to meaningful overarching values and goals, they take on different meanings. Sweeping the floor can be transformed from a boring task that is resented to a powerful means of serving others. One recent study demonstrated that workers were willing to be paid 44 percent less for the same job once they learned that their prospective employers had a social responsibility mission statement.[11] Another experimental study showed that workers who were matched with employers who shared their personal missions produced 72 percent more output than unmatched workers. In other words, the effort of one matched worker was equivalent to approximately 1.7 unmatched workers.[12] Interestingly, the positive effects of being well aligned with the purpose of the group to which they were assigned was much greater than the effects of offering large extrinsic incentives.

Shared purpose facilitates cooperation with other groups. A group that is clear about its purpose can telegraph its intentions clearly to others, building trust and making the group more attractive as a partner for collaboration. When a group is clear about what it does and why, other groups can more easily engage and relate to the group.

Shared purpose creates personal benefits. Connecting and reconnecting with purpose also has personal benefits. For example, it helps make people more resilient to stress. One study created a stressful situation by asking participants to give an impromptu speech. Before the speech, all participants were asked to rank their values and describe issues that were important to them, but a subset was also asked to talk about the personal value that most mattered to them. These participants had a lowered cortisol response to the induced stress, which buffered its damaging effects.[13]

Connecting with purpose and aspirations can even increase a person's status in a group. One fascinating study asked participants to write a couple of paragraphs describing their aspirations. Another group was asked to write about the duties and obligations they had to fulfill in life to avoid disapproval. And another "neutral" group was asked to simply describe their commute to campus that day. In the language of the matrix, the first condition primed people to think about what they wanted more of in their lives, or "toward" thinking, whereas the second condition tended to prime people for prevention, or "away" thinking, because their writing prompt emphasized activities they felt compelled to do by others. Participants were then asked to perform a ranking task in a group with people they had not met before. Participants in the "toward" condition were ranked as being more respected, admired, and influential by their teammates than those in the "away" and neutral conditions. The results show that when people are primed to focus on approach rather than avoidance, they are more proactive in contributing effort and ideas to the group, as well as in managing their own needs.[14]

Shared purpose makes collective interests more salient and highly valued. Perhaps most importantly in terms of our and Ostrom's analysis, working to create shared purpose *is* the very essence of clarifying and prioritizing the collective interests of the group over individual interests. When the purposes of individuals within the group do not align with those of the group as a whole, a lot of time and energy is spent attempting to control individualistic behaviors. If a group can bring individual and collective purposes into alignment, a lot of this time and energy can be devoted to finding ways to better cooperate rather than to control individualism.

So, now you know why shared purpose is particularly important both to create shared identity and to coordinate activity. Now, how do you ensure your group has shared purpose? And once you do, how do you make sure your group is sticking to it—that your stated shared purpose truly functions as a guide to your group's action? We'll discuss this for the remainder of the chapter.

Mapping Collective Interests with the Collective Matrix

The collective matrix is a superb tool for clarifying and deepening a sense of shared purpose in most groups. Not only can it help establish values and goals, but the process of discussing the fears and concerns that surface and might get in the way can often help with the process of integrating individual and collective interests, while building a sense of trust and safety in the group.

Instead of beginning the process with "What matters most to *me* about the group?" you can begin with questions like "What is *our* shared purpose? Who or what matters to *us*?" You can then follow up with questions similar to those we outlined with the individual matrix exercise, but framed as a collective (see figure 7.1).

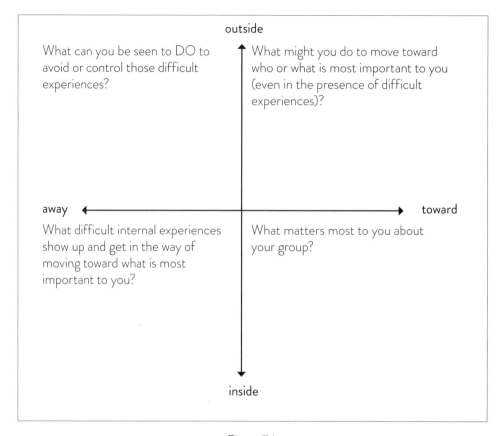

Figure 7.1

If you've done the individual matrix in depth using a digital tool, such as one of the ones recommended at http://www.prosocial.world/digital-tools, you may find that there's some redundancy in the responses on the left-hand side of the matrix. But there will also be many differences, because people are now reflecting on what gets in the way of them moving fully toward the group's goals and values, not their personal goals and values.

Typically when a group uses the collective matrix to create shared identity and purpose, there will usually be a point when you need to group the similar responses on the right-hand side and prioritize them. Both these processes can be easily implemented using sticky notes, or online using a tool (see http://www.prosocial.world/digital-tools) that allows for grouping and voting.

To give you an example of what a completed matrix might look like, we've included examples from a collective matrix Paul did with one of his Prosocial facilitator-training courses, in which he started with these questions: Who or what matters most to us (as a Prosocial community of practice)? What is our shared purpose? (See figure 7.2 for the results.) For each quadrant they first brainstormed ideas and then organized them into meaningful groups of statements. The process took less than forty-five minutes and greatly enriched the group's clarity of purpose and, just as importantly, its shared sense of trust and identity—so much so that this map formed the basis of their work for the rest of the class and beyond.

Core Design Principle 1: Shared Identity and Purpose

What might WE be seen doing to move away from that bottom-left stuff?

- Occupy ourselves with other stuff
- Proceeding step-by-step
- Keeping quiet
- Not visiting community site
- Limiting access, using rigid copyrights
- Not sharing material
- Taking without contributing
- Pretending to be interested
- Splitting up into groups
- Avoiding hard conversations (decision making, what is Prosocial)
- Getting protective, not sharing
- Not actively marketing it
- Not doing Prosocial for our own groups of groups
- Using buzzwords just to please clients
- Keeping others out, engaging only with the "good guys"
- Fantasizing instead of acting
- Selecting groups to do Prosocial that are already doing well
- Fighting about control, competing and discrediting instead of collaborating

What could WE be seen doing to move toward our shared purposes (even in the presence of bottom-left stuff)?

- A bank of shared resources
- Using a creative commons license to share resources
- Going to website actively (community)
- Celebrating successes
- Working in teams
- Supporting each other (e.g., generously offering advice, sharing experiences, speaking up, discussions)
- Telling others outside of the community about Prosocial
- Sharing challenges and learning
- Promoting Prosocial in organizations
- Opening conversations about bottom-left stuff
- Offering Prosocial/materials to groups who can't afford it
- Peer-review process for Prosocial facilitation
- Talking about Prosocial at a high level of hierarchy in organizations
- Prosocial meetups, monthly hangouts
- Establishing clear roles and guidelines for who does what in the community

← away   outside   toward →
                   inside

What might show up inside US (not just you but others as well) that might hook us and get in the way of us moving toward our shared goals?

- "I have no time"
- INTERNAL POWER STRUGGLES
- "Who has the right to influence what, how, and how much? And who decides this anyway?"
- SELFISHNESS
- FEAR of showing vulnerability
- "we know it better than others"
- "It's too big"
- DOUBT
- "I'm not expert enough"
- ANXIETY and FEAR about goals of other groups
- "What, I'm putting all my little time into this and people pick it for free?"
- "Others will take our ideas and present them as theirs"
- "This will not be good enough to share"
- "Too academic, no real outcome for business"
- DOUBT about control of the community
- "I won't get enough time with groups"
- "I could do better doing it myself"
- "I don't know where to start"
- "They will not take it seriously"
- "People will NOT get it; they're only interested in the next best quick fix"

Who or what matters most to US (as a Prosocial community of practice)? What is our shared purpose?

- Sharing material
- Sharing best practices
- Collaborative spirit
- Openness
- Changing the world together
- Increasing well-being
- More love, less hate
- Values-focused
- Creating healthier and more harmonious groups
- Helping groups work more meaningfully towards their shared purposes
- Helping people grow
- Promotion of polycentric governance
- An evidence-based approach
- A science-based approach
- Integrity
- Making the world more beautiful
- Improving well-being
- Getting inspired
- A new balance between organizations, communities, political groups
- Support and supervision
- Ownership by the community
- Making Prosocial accessible to those who need it
- Empowering people, without need for resources
- Organizations where people feel respected and connected and like to work

Figure 7.2

You can play with the order in which you complete the quadrants to suit your context. In the work that Paul did with the ACT trainers, he and his cofacilitator Beate Ebert decided to begin in the top-right quadrant because the most salient thing for the group was what they wanted to be doing more of. Figure 7.3 includes the questions they asked, numbered in the order they asked them.

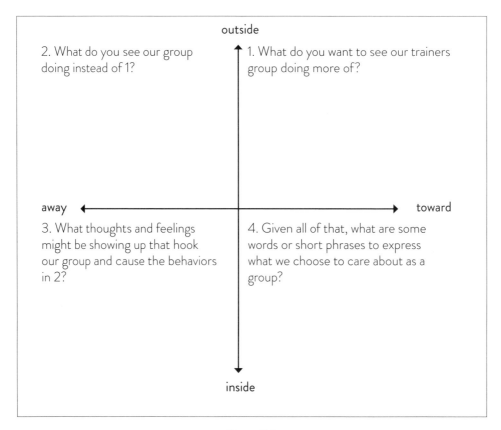

Figure 7.3

Perspective Taking

If your group seems to struggle with the question "What is our shared purpose?" you can get them started with other approaches. For example, in the Prosocial process we often deliberately shift a group's perspective to a future time frame. For example, we might ask this: Imagine it's five years in the future, the group has been extraordinarily effective. What would you be seeing and noticing? What would the group have achieved? This line of questioning moves group members forward in time, drawing on the flexible perspective-taking skill of "now/then." Alternatively, you can use the other distinctions from which the human sense of self arises ("I/you" and "here/there") by, for example, asking participants to imagine how others might see their purpose.

You can also ask questions that move people backward in time: What has enlivened and excited us in the past? When are we at our best? This can spur group members to evaluate what they're doing now and how it bears on the actions of the group as a whole, which leads members to consider what the group currently is, what they feel it should be, and what they want it to be.

Moving from How and What to Why

It's very easy for a group to mistake the *how* and the *what* of what it does for its purpose—that is, the *why* it does what it does. This happens all the time. As soon as there are processes in groups, people mistake those processes for the reason the group exists. For example, in a company the accounting group is devoted to accounting, the HR group is devoted to enforcing HR rules, and so on, and as the individual groups plug away at what they do, the collective purpose of the organization is lost. The purpose no longer animates and enlivens the work individual employees do; it no longer gives their work meaning or encourages them to behave prosocially.

If this is a problem for your group, it's helpful to learn to "ladder up" and "ladder down" the hierarchy of purposes to find one that is sufficiently general enough that everyone can agree on it. For example, people might disagree about specific ways to save the biosphere, but it's likely that they all agree on the importance of doing so. Laddering up and down is also useful for finding a purpose that is maximally motivating while still being specific enough to provide guidance.

One simple approach to laddering is to use a series of "why" questions to get more general and a series of "how" questions to get more specific. For example, Paul worked with a local environmental action group that expressed its purpose in a range of ways, from "saving the planet" down to "coordinating bulk buys of solar panels." By asking "Why does that matter?" to move up the hierarchy, and "How would you do that?" to move down it, the group was able to settle on an intermediate purpose that motivated the group: guiding and inspiring local action to improve our environment. Here are some examples of questions that can help you move down the hierarchy:

What could you do to contribute to this goal or quality of action?

A year from now, what would you like to have accomplished in the service of the overarching goal of X?

What do other successful teams do in the service of their goals?[15]

Getting the Level Right

The main aims of having a clear sense of purpose are to guide behavior and motivate action. Sometimes it can feel like there's tension between these aims. Specific purposes provide clear guidance on behavior but sometimes do little to motivate group members. On the other hand, grand purposes do a great job of motivating but offer little guidance on behavior. The key is to find a balance. The purpose of a community group teaching salsa is surely to teach the dance technique—clearly, a guide to the way its members should behave. But reflecting on the deeper purpose of the salsa group might reveal that its work is just as much about building community, providing an opportunity for mixing fun with exercise, or even teaching people about Spanish culture.

Conversely, Steve Jobs is often quoted as having said his purpose for himself and Apple was to "put a dent in the universe." The grandeur of the metaphor acts in a sort of meta way, encouraging individuals to construct more specific purposes that are rich, grand, and perhaps riskier than they otherwise would have been, but this purpose says nothing about what Jobs wanted members of the Apple corporation to actually do. When developing a shared purpose, pay attention to whether group members are both excited by the purpose and able to use it as a guide to influence their decision making. Also, keep in mind that not all group purposes are the same. Purposes like "being the best" or "dominating the market" are more or less meaningless; for most, such purposes simply express a desire to win out over competitors, or to only benefit those in charge rather than the group as a whole.

Ultimately, the most meaningful and powerful purposes are those that involve making a positive contribution to the world. Fund-raisers for scholarships feel more motivated when they can talk with the recipients of those scholarships;[16] chefs work harder when they see the people who will be eating their food.[17] Just watching a short video showing examples of people making a positive impact on others within a tomato-processing company led to a 7 percent increase in tomato harvesting, and this increase was sustained for weeks.[18] Helping others increases feelings of belonging because people feel valued by and connected to those they help.[19]

Now, it is not enough to just set the purpose and move on. Too often that leads to people forgetting the purpose. Your group's members will need to *live* the purpose—by keeping their individual expressions of the collective purpose in mind in their day-to-day work and in the ways they make their decisions as group members (core design principle 3), monitor themselves and others (core design principle 4), address their conflicts (core design principle 6), and more. It's this living, dynamic aspect of purpose that we explore next.

Living Purpose

The simplest and most important way to build a workable sense of shared purpose in a group is simply to encourage members to continually ask questions like "What matters most in this situation?" and "What are we trying to achieve here?" As often as not, purpose emerges from the work of the group, as illustrated by the following story.

Joan was at the front desk of a community library in Melbourne when she looked up from her work to see ten Afghanis standing at the counter. They asked her if they could join the library. Joan knew that, under existing library policy, refugees without permanent visa status were not allowed to join, but she also knew, looking at her customers, that this policy was not right. Asking the refugees to come back in a week, she went to her boss and then her boss's boss all the way up the line to the head of the library, who agreed with Joan. They managed to get the policy changed, and this story then became a central piece of the library's new purpose: access for all. The purpose arose from paying attention to what really matters.

Purposes are "lived in" when they both grow from and inform real decision making, such as where to allocate resources or whom to hire (core design principle 3, fair and inclusive decision making). When a group deliberately selects members based on their alignment with the group's purpose, the group is more likely to build shared identity through prosocial behavior that enhances group safety and performance. The purpose must also inform processes of performance management (core design principles 4 and 5, monitoring agreed behaviors and graduated responding to helpful and unhelpful behavior, respectively), the resolution of differences (core design principle 6, fast and fair conflict resolution), and relations with other groups (core design principles 7 and 8, authority to self-govern and collaborative relations with other groups, respectively). Some innovative groups leave an empty chair in the room to symbolize the collective purpose of the group while making decisions. Having a seat in the conversation for "us" or "we" makes the idea of a collective interest more real and present. If it weren't such a clunky way of saying it, we would have preferred to have called principle 1 "Shared identifying and purposing in context" to emphasize its active and contextually situated qualities.

Purposes become even more lived in when they are used to make hard choices. One research group decided to turn down funding from a company that built and sold poker machines because the research group believed the aims of the poker machine company conflicted with its purpose of improving societal well-being. Another group had to make the hard choice to lay off a senior member of its team when it became clear that his values and beliefs conflicted with the stated purpose of the group to increase diversity and inclusion. While these decisions were painful, they did more than anything else to embody and authenticate the purposes of these groups.

In a group that lives and breathes purpose, where the purpose is embedded in every conversation, every agreement, and every process, *everything* is evaluated against that purpose. One survey showed that more than half (53 percent) of change efforts succeeded at companies with a strong sense of purpose, whereas less than one-fifth (19 percent) succeeded at companies without a shared sense of purpose.[20] Consistent with this finding, 86 percent of executives surveyed believed that shared purpose was important, but fewer than half (46 percent) of the companies surveyed had a clear sense of purpose.[21] Hopefully this chapter has demonstrated that finding shared identity and purpose is at the very heart of group cooperation and well worth focusing on using the methods we've described.

Conclusion

In this chapter we briefly surveyed why shared identity and purpose are important for groups and offered some ideas for improving both. We focused on the collective matrix as the primary tool for building shared identity and purpose, but of course there are many other methods you can use.[22] The most important thing to keep in mind with core design principle 1 is to continually revisit and use the group's purpose in every aspect of its day-to-day work, including for making decisions. In doing so, the capacity of the group's purpose to both guide and motivate will continue to deepen.

In the next chapter we turn to the second core design principle: equitable distribution of contributions and benefits.

CHAPTER 8

Core Design Principle 2
Equitable Distribution of Contributions and Benefits

Have you ever been part of a group in which workloads were distributed unevenly or some members were unfairly given more recognition than others or received more perks, such as better promotional prospects, more training, or increased contact with senior members of the organization? If so, you know how perceptions of unfairness can destroy cooperation. Or perhaps you can recall a time when you were treated fairly, when someone went out of their way to include you in a decision that affected you, or there was a transparent process of decision making based on clear principles. Even if the decision went against you, it's likely that you cooperated with the outcome and felt satisfied by your actions.

Talking about fairness is typically hard for individuals and groups. People worry about sounding too demanding, coming across as selfish, being unkind to others, or escalating conflict. But fairness, or the equitable distribution of contributions and benefits (core design principle 2), is key to having a highly functioning group.

Fairness is just as important, or perhaps even more important, *between* groups as it is within groups. It's easy for one group to feel as though another group is benefiting more than is warranted by its contributions. The daily media is full of discussions about whether different nations are acting fairly in relation to such issues as trade or climate change, and the same issues play out in strained relations between various branches of local, state, and federal governments. Even within organizations it's common for some teams to feel as though another team receives unfair advantages.

In this chapter, we'll first describe one case in which fairness really mattered. We'll then briefly review what fairness means in the context of groups and why it is so important. Then we'll discuss how we decide whether groups need to be talking more about fairness. Some groups manage fairness just fine as a kind of side effect of focusing on purpose, inclusive decision making, transparency, responsiveness, conflict resolution, and leadership. But for some groups, fairness is *the* primary issue they're contending with, and having the space to talk about it cuts through the issue like no single conversation can. Because of the importance of creating safe and productive conversations about fairness in groups, we devote most of this chapter to describing how we approach doing so.

The Research Institute

Paul was asked to conduct the Prosocial process at a highly esteemed research institute. The institute prided itself on attracting significant grant funding, publishing in top journals, attracting high-quality research staff, and having an impact on society. Like most academic institutions, employees were broadly grouped as tenured academic staff responsible for conducting research, and as professional staff responsible for conducting the day-to-day running of the institute and for supporting the academics. Academics were evaluated by their publications and success in attracting grant funding. Professional staff members, both administrative and research support, were primarily evaluated based on their responsiveness, effectiveness, and efficiency.

While the institute was performing well on academic measures, the professional staff members were unhappy, resulting in high turnover and frequent complaints. Paul conducted a survey using the Prosocial design principles. He found that professional staff members rated the organization much lower than the academic staff on almost all of the core design principles (see figure 8.1).

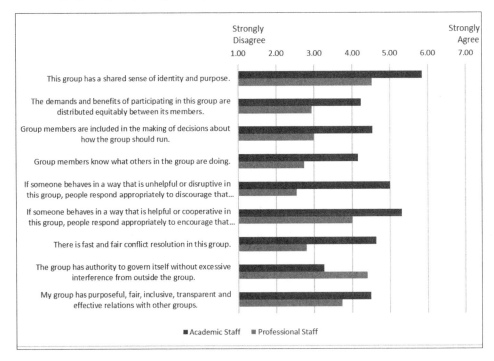

Figure 8.1. Academic and professional staff ratings of performance based on the core design principles.

Many of the issues that professional staff members raised had to do with fairness. For example, they frequently were not involved in making decisions about how projects should run or who should be involved, even though their performance was markedly affected by these decisions. And the names of a subgroup of younger professional staff members, who had lots of research expertise and aspired to be academics, were usually not included on publications despite them having contributed substantially to the research. Also, unlike with academic staff members, there were few promotional opportunities for most professional staff members, and many left complaining about the lack of a career path. Something had to change.

The Importance of Fairness

Fairness matters a lot to people within and between groups. It's built into human prosociality, not to mention human nature. If we care about others, and if we have any capacity to take the perspective of others, we invariably compare our own contributions and outcomes to those of others. This sensitivity to fairness arises early in human development, around the same time as language acquisition.[1] As a result, it appears to be innate and thus has been selected for over time. We are continuously on the lookout for how we are doing relative to others and for making sure we're not exploited.[2] Fairness is the very essence of balancing interests: "my" interests with "our" interests within groups, and "our" interests with "their" interests between groups.

To care about fairness—to care about others and to step into their perspective to ensure that their needs, too, have been taken care of—is to be prosocial. Indeed, under the right conditions, people will quite happily act *against* their own self-interest to restore a sense of fairness, even if the relationship is temporary or anonymous.[3] Literally thousands of studies over decades of research have shown that people will readily sacrifice benefits for themselves to distribute resources fairly.[4]

Given the extent to which a sensitivity to fairness is woven into our psyches, unfairness can be incredibly toxic within and between groups, almost invariably reducing performance and causing individuals or whole groups to exert less discretionary and cooperative effort. Just as we saw illustrated in the research institute case, groups that suffer from inequity have increased negative attitudes, reduced commitment to group goals, and higher rates of turnover.[5] People in such groups are more likely to be stressed and unhappy and are more likely to engage in destructive behaviors, such as theft, retribution, revenge, sabotage, and retaliation.[6] In one study, two companies both implemented a 15 percent pay cut. During an extended meeting with staff, the president of one company fully explained the rationale for the cut. With the other company, the vice president briefly explained the cut in a ten-minute-long meeting. Over the ten weeks following the pay cut, theft was 80 percent lower in the first firm than the second one, and employees were 15 times less likely to resign.[7]

On the flip side, when employees believe that organizational resources are allocated fairly, that processes for decision making are fair, and that the organization treats them fairly, they are more motivated. They're more likely to cooperate, offer assistance freely to others, and support the decisions of the group.[8] And they are more likely to develop close relationships with others in the organization. They are also more likely to be creative and happy. One study even showed that training supervisors in methods of fairness led to a decrease in insomnia among the nurses they supervised. Nurses were literally losing sleep over their feelings of injustice.[9]

People who feel they have been treated fairly by their organization are more likely to internalize the organization's goals and values and to feel more included and respected, thereby increasing a sense of shared identity with the group.[10] Fairness increases trust and reduces perceived uncertainty.[11] Even in large groups, people are more likely to act prosocially when they are treated fairly. For example, during the 1991 water shortage in California, people saved more water if they felt the authorities had treated them fairly.[12] And in another study, people were more willing to make contributions to municipal child care if they felt the public good was distributed fairly.[13]

So if group members believe they are being treated unfairly, the group cannot thrive. This is just as true for relations between groups as it is for relations between people. How, then, can you best work with core design principle 2 to increase perceptions of fairness? It is to that question that we now turn.

Working with Core Design Principle 2

"Fairness" is a vague term, and one of the first steps to working with groups on this topic is to pin down more precise ways of talking about it. The first really useful distinction to make is between distributive and procedural forms of fairness. *Distributive fairness*, called distributive justice in the research literature, is about how resources are distributed among people—for example, who gets how much money, how much training, or promotional opportunities. *Procedural fairness* is about whether fair processes occur. Are people involved appropriately in decisions that affect them, is there transparency, and so forth? Core design principle 2 is focused mainly on distributive fairness, while subsequent principles, particularly core design principles 3 (fair and inclusive decision making) and 6 (fast and fair conflict resolution), are focused on procedural fairness. In this chapter we'll focus mainly on the distributive aspects of fairness. Furthermore, although we'll focus primarily on within-group examples, please remember that the power of these principles is their capacity to be scaled up to the groups of groups level, so much of what we say could be equally applied to that level.

Focusing Conversation in a Helpful Way

Once you focus on thinking about distributive fairness, the next useful distinction to be aware of is norms of fairness. Consider this example: Three employees of a medical goods company, Alison, John, and Margaret, are arguing about who should be given the perk of an overseas sales-training program. Alison argues that she made the largest number of sales this year and deserves to be rewarded. John argues that Alison previously received a similar trip and that it's his turn. Margaret argues that she really needs the training as the lack of it has been holding her back from making as many sales as Alison.

These three positions represent three different ethical norms of fairness, which are useful to understand when facilitating groups. Alison's approach, an *equity approach*, takes merit and contribution into account when deciding on the distribution of resources. By contrast, John's approach, the *equality approach*, provides equal benefits irrespective of contribution. Margaret's approach, *allocation according to need*, is a third normative principle that strives to achieve equal final outcomes for all, so that those who are more disadvantaged to begin with receive more irrespective of contribution.[14]

When group goals focus on high performance, groups tend to value a norm of equity. But when they are focused on social harmony and care, they tend to value norms of equality or need. Each approach has costs and benefits. Equity motivates highly skilled individuals to contribute as much as possible to the group, but if handled poorly it can create competition and comparison between individuals, thereby reducing overall group cohesion. Equality emphasizes the equal worth of all but can create situations in which people free-ride on the efforts of others. Need emphasizes a principle of care for all but can create feelings of resentment when those who contribute little are sometimes carried by those who contribute a great deal.[15]

Of course, these norms are not mutually exclusive, and some groups employ two or even all three of them at different times in different contexts. For example, Alison might be given recognition in the staff newsletter, and Margaret might get the overseas training she needs to improve her sales, thereby recognizing both high performance and need. The important point here is not to favor one approach or the other, but to be aware of when different norms are being applied. In our experience, much of the conflict in groups that comes from unfairness results from not recognizing when people are applying different standards for what is fair.

While we are strong believers in the passion, motivation, and commitment that can be generated when benefits flow to people who devote time, effort, and energy to contributing to a group (equity principle), we are also highly aware that there is not always a level playing field. For example, we have seen many groups in which hierarchical leaders received disproportionate benefits because they had a greater impact on the functioning of the group—but their greater contributions were not a result of greater effort or commitment, but of greater access to resources, influence, and visibility. From our perspective, the most effective groups begin with a recognition that everybody's interests are inherently just as important as everybody else's, and everybody deserves the opportunity to contribute to the fullest extent possible, if they so desire. The ethic of equity held rigidly can become a recipe for "the rich get richer" and must always be tempered with equality (at least in the most basic sense that every human's needs are as important as any others) and need (at least in the sense of access to opportunities to contribute). While it might be procedurally fair to state up front that anybody who makes additional contributions will be rewarded, everybody must have access to the opportunity to contribute.[16]

You get a particularly clear view of these norms at play when you look across cultures. Some cultures emphasize equality; others prefer equity. In some cultures, receiving too much is a cause of as much distress as receiving too little, while in others gaining advantage from an exchange is quite acceptable.[17] The specific details of what we call equity are not something we're born with; equity is culturally constructed in our language, our values, and our social comparisons.

These different norms informed the label we chose for this principle. We used the term "distribution" because we wanted to flag this principle as being primarily about distributive fairness. We focused on being "equitable" because many of the groups we work with value quality of cooperation and performance over harmony. By "contribution" we mean any form of giving to the good of the group, perhaps through exerting additional effort, working longer hours, or contributing resources. In practice, the most frequent contribution a person makes to a group is to spend time working for the good of the group. By "benefits" we mean anything that benefits the individual. So, benefits might include salary, public recognition, learning opportunities, or just social engagement.

Although we focus on equity as the primary distributive norm in the groups we encounter, we recognize that groups in particular contexts, such as an intentional community project, a community-based arts program, or a national health service, may decide that distributing resources equally independent of contribution, or distributing resources according to need rather than contribution, is fairer. For example, a not-for-profit group we worked with recognized that, both within its team and among its clients, people start with very different levels of resources and capabilities to contribute. For this group, the ethic of equality captured something deeply important to it, and the group felt as though it was living more closely to its values when benefits and workload were assigned equally. Other groups may come to a different conclusion. For example, high-performing teams in commercial organizations may decide that benefits should be directly related to performance (an equity ethic) to reward discretionary effort and skill development. In this sense, discussions of fairness draw upon discussions about shared purpose and values (core design principle 1, shared identity and purpose), and thus there is a natural ordering to the principles.

So, once the ideas of distributive and procedural fairness, and the differences between equity, equality, and need-based norms of fairness, are introduced to groups, they are usually in a much better position to enrich their discussions of what really matters in relation to principle 2, and to link issues of fairness to their shared purpose.

It might seem that the next obvious step would be to ask, "How fairly do you think you have been treated in this group?" This question is fine when the group is doing well and fairness is not a problem, but for groups in which it *is* a problem, in our experience this direct approach is often unhelpful. For example, with the research institute, asking directly about fairness tended to evoke a string of rigid, oft-repeated stories and complaints from those who felt unfairly treated, as well as a string of justifications for current policies from those who were seen as benefiting. In the next section we explore why this question is often unhelpful, and what you might do instead.

Don't Focus on How Fairly People Believe They've Been Treated!

If fairness is a problem for a group, it has been our experience that opening a conversation focused on fairness can often backfire. Why? We see several problems with this focus, including the psychological inflexibility of members and that it primes self-interest and encourages biased and inaccurate estimates of contributions and benefits.

Not being treated fairly is painful, and when that pain is wrapped up in strong judgment you have a formula for psychological inflexibility. If you have children, you know that crying in the back seat of the car is often associated with the words "That's not fair!" Adults are not immune to that same pull once judgments are dominant.

Focusing on "fairness" is likely to lead to such judgments because it causes group members to zero in on how much they are getting out of the group, rather than on their shared purpose. In this sense, the question can sometimes redirect attention toward "What's in it for me?" instead of "What's in it for us?"

People are notoriously self-serving when comparing their contributions and benefits to those of others. It's much easier to see one's own contributions than it is so see those of others. If you ask the partners in a married couple each to separately estimate what proportion of the chores they do at home, the total of the two will usually considerably exceed 100 percent! Similarly, supporters of one sports team give lower estimates of the number of fouls committed by their team than do supporters of other teams.

Furthermore, people can always disagree about fairness simply by changing the baseline that they choose for comparison. For example, to prove that I am being treated unfairly I might choose to compare my salary outcomes to highly qualified people. Meanwhile, to prove I am wrong, my manager might choose to compare my salary to that of a new graduate. Who's right?

This changing of baselines can even extend to choosing completely different ethical norms. People who spend more time on a task believe they should be paid more according to their inputs (an equity rule), while those who work less believe they should paid the same amount as those who worked more (an equality rule). All of these examples were cited in one classic study that showed that in a mock court case the views of participants regarding what would be a "fair" outcome were automatically distorted to be in line with whatever role they had been randomly assigned in the case. They were completely unaware that they were biased, and they were given incentives to be as accurate as possible![18]

There are many other ways that people are biased in their estimations of fairness. People and groups who have more resources are much more likely to argue for an equity-based model of fairness, whereas people and groups who have fewer resources are much more likely to argue for an equality- or need-based system of allocation. And powerful people have stronger expectations that they will be treated fairly, and they react more quickly to violations of fairness.[19] All of these patterns make sense from a multilevel evolutionary perspective, which explains both our capacity to cooperate and our capacity to cheat!

As with a lot of issues in groups, problems arise when the focus is on the outcome, not the process that creates the outcome. Just as we cannot force love or sleep, we need to allow fairness to emerge from conversations about individual and collective needs and interests. These need to be conversations in which people are willing to talk about what they care about, experience discomfort, stay engaged, actively step into the worldviews of others, and acknowledge the automatic influences of self-serving biases.

In summary, starting a conversation by asking about fairness directly is risky. At the same time, not talking about unfairness can a kill group's vitality or even tear it apart. What are we to do then? In the next section, we explore another more productive way into talking about the issue of equitable distribution of contributions and benefits.

Using the Collective Matrix to Improve Fairness (Flexibly)

In our experience, getting out of conversations focused on a win-lose transactional form of fairness, and what members feel they deserve, is critical. Improving a sense of fairness in a group results from the search for mutual needs satisfaction, mutual perspective taking, and mutual forgiveness for momentary or minor transgressions—in essence, conversations that happen in a win-win space. Fairness grows from a group culture that's focused on "us" (that is, shared purpose and identity, core design principle 1) rather than one focused on me comparing my outcomes to your outcomes.

The individual and collective matrices are enormously helpful with this because they help build shared purpose and trust. When people are focused on their bigger and shared "toward" moves, there's less emphasis on the more self-protective "away" moves that are involved in social comparison. A conversation focused on "Are *we* working together effectively to achieve our shared purpose" is more likely to produce fairness as a by-product than a conversation focused directly on "Am *I* getting what I deserve?" This is particularly true when the dominant norm for fairness is equity—that is, benefits relative to contributions. Only with a clear sense of shared purpose can there be clarity and agreement about what constitutes a valuable contribution.

Let's imagine you are working with a group for whom distributive fairness seems to be a particular issue. You might have already done the individual matrix and discovered that a perceived lack of distributive fairness is commonly mentioned on the left-hand side. Or perhaps you've done the Prosocial survey with your group and discovered that core design principle 2 is rated as being poorly implemented. How might you use the collective matrix to help the group discuss what's going on and plan a way forward?

Start with the Bottom-Right Quadrant: Values

Let's imagine you've called your group together and have decided to use the collective matrix to help its members talk about the issue. This is not what we did with the research institute because the issues of inequity seemed to be related to problems with a number of the core design principles, so we decided to focus on multiple principles at the same time. However, for the purposes of illustration, from that example we pulled together the aspects of the process that were specifically focused on fairness.

We might, for example, introduce the session as follows:

1. Present the evidence showing there's perceived inequity (for example, the Prosocial survey, results from the individual matrix work, or just informal conversations among staff).

2. Remind the group of its shared purpose (from earlier work, perhaps with the collective matrix) and the importance of taking care of everybody's needs and values in order to achieve that shared purpose.

3. If you think it might be helpful, either introduce some process guidelines (that is, ground rules) for the conversation or, better still, if you have the time, ask the group, "How should we have this conversation to ensure that people feel it's productive and safe enough to contribute?" Usually people come up with ideas like listen carefully and reflect back what people have said and separate facts from evaluations.[20]

4. Start the bottom-right quadrant with a question like "What matters most to us about this situation?"[21] To go back to our example of the research institute, professional staff members felt that they were treated unfairly not just because the outcomes they received seemed unfair relative to their contributions, but also because they perceived that they were treated with a lack of respect. It is possible to reframe these complaints as values by simply turning them on their head. In this case, we might reasonably suppose that values of fairness, equity, and respect are important to this group (see figure 8.2). As you can see, in the context of a group with fairness issues, this question usually evokes a rich tapestry of values related to fairness.

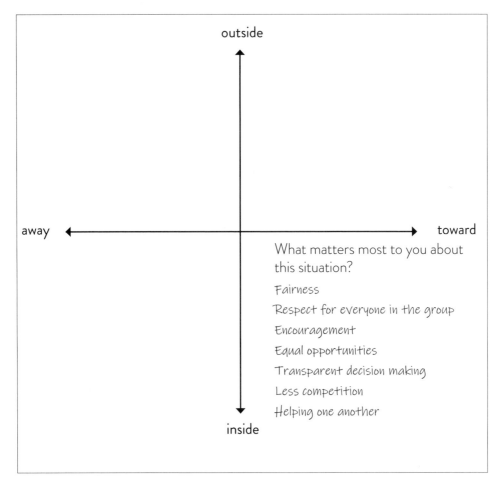

Figure 8.2

At this point with the collective matrix, you always have a choice about which quadrant to move to next. Typically, you'll want to move to either the top right (goals and concrete behaviors that will move the group toward its values) or the bottom left (hooks that get in the way of moving toward values). It can be helpful to move to the top right if you think the group would benefit from getting more concrete about what fairness actually would look like in the group. It can be helpful to move to the bottom left if the difficult thoughts and emotions that arise in the context of unfairness are more pressing. For our illustration, we'll go to the top right to get specific about what greater fairness might look like.

Move to the Top-Right Quadrant: Preliminary Ideas About How Fairness Could Be Improved

You could introduce the top-right quadrant by saying something like, "Okay, we've talked about what is most important to us in this situation. Let's talk now about some preliminary ideas for what we need to do as a group that would improve the situation." The group might then brainstorm a range of options (see figure 8.3).

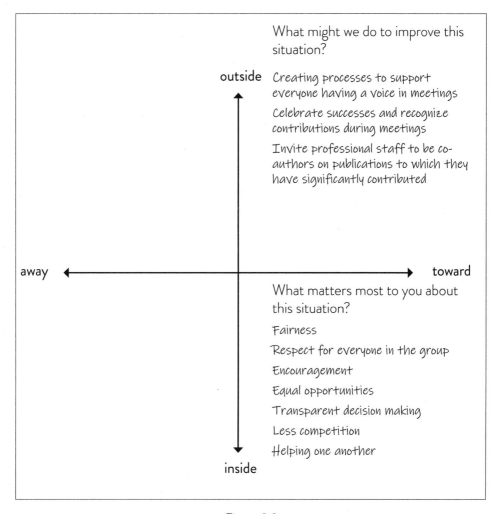

Figure 8.3

Then, once you have some initial specific suggestions for what might actually be done to address issues of fairness, you can move down to the bottom-left quadrant to discuss the hooks that come up when considering all the material on the right-hand side.

Move to the Bottom-Left Quadrant: Hooks That Get in the Way of Group Fairness

Let's now move to the bottom-left quadrant of the matrix. What internal experiences typically show up for people and hook them into acting in ways that work against the material on the "toward" side of the matrix? Figure 8.4 shows the kinds of material we heard from staff at the research institute when we spoke to them confidentially about the proposed changes on the top-right of the matrix.

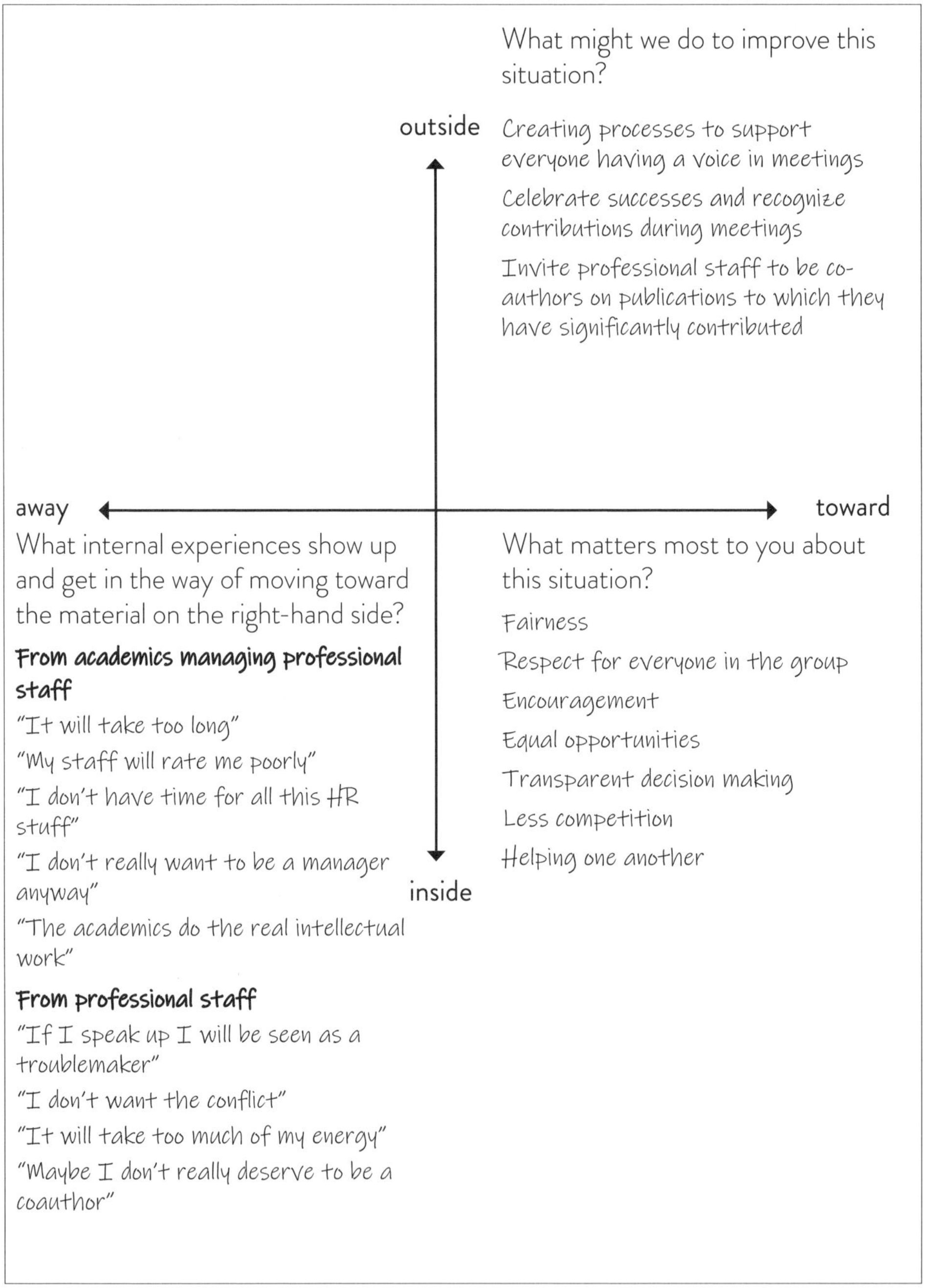

	outside ↑
	What might we do to improve this situation? Creating processes to support everyone having a voice in meetings Celebrate successes and recognize contributions during meetings Invite professional staff to be co-authors on publications to which they have significantly contributed

away ←————————————————→ toward

What internal experiences show up and get in the way of moving toward the material on the right-hand side?

From academics managing professional staff
"It will take too long"
"My staff will rate me poorly"
"I don't have time for all this HR stuff"
"I don't really want to be a manager anyway"
"The academics do the real intellectual work"

From professional staff
"If I speak up I will be seen as a troublemaker"
"I don't want the conflict"
"It will take too much of my energy"
"Maybe I don't really deserve to be a coauthor"

What matters most to you about this situation?
Fairness
Respect for everyone in the group
Encouragement
Equal opportunities
Transparent decision making
Less competition
Helping one another

↓ inside

Figure 8.4

We go out of our way to remind groups that the sort of self-protective responding we often see in the bottom right is perfectly natural in the sense that we have evolved to protect ourselves as well as to cooperate in groups. If you are to use the matrix to build trust in groups, it's important to find ways to make it okay to be honest about left-hand-side material in the matrix. As we've mentioned, you might do this by teaching effective listening skills, conducting prework focused on creating more shared purpose and identity, or using tools that support anonymous reporting.

Move to the Top-Left Quadrant: Defensive Routines

So, what do groups do when they're in the grip of fear and other hooks, such as those described in the previous section? The most common reaction we've seen is that they simply avoid altogether addressing issues of perceived unfairness. Group members collude in this avoidance for fear of creating conflicts that cannot be resolved. In the absence of trust in the other core design principles and the psychological flexibility of those in charge, it is deeply threatening to raise issues of inequity and unfairness. Over time everybody learns to just put up with the unfairness. Another common reaction is to resort to formal policies and procedures, such as diversity or grievance policies, and claiming that these ensure fairness when of course they are often divorced from context and do nothing to address the fairness concerns of the group.

As we've already discussed, prolonged unfairness tends to lead to less discretionary and cooperative effort, increased gossiping about how bad the group is, and maybe even group members choosing to leave. In extreme cases, people sometimes react to unfairness by engaging in destructive behaviors, such as theft, retribution, revenge, sabotage, and retaliation.[22] So now our collective matrix might look something like figure 8.5.

What can you be seen to DO to avoid or control those difficult experiences? Have unstructured meetings Fail to discuss the issue Rely upon HR to tell us what we need to do in relation to staff Only celebrate staff achievements when they have already received recognition by others Give up and just put up with it Withdraw discretionary effort	**What might we do to improve this situation?** Creating processes to support everyone having a voice in meetings Celebrate successes and recognize contributions during meetings Invite professional staff to be co-authors on publications to which they have significantly contributed
What internal experiences show up and get in the way of moving toward the material on the right-hand side? **From academics managing professional staff** "It will take too long" "My staff will rate me poorly" "I don't have time for all this HR stuff" "I don't really want to be a manager anyway" "The academics do the real intellectual work" **From professional staff** "If I speak up I will be seen as a troublemaker" "I don't want the conflict" "It will take too much of my energy" "Maybe I don't really deserve to be a coauthor"	**What matters most to you about this situation?** Fairness Respect for everyone in the group Encouragement Equal opportunities Transparent decision making Less competition Helping one another

Axes: outside / inside (vertical), away / toward (horizontal)

Figure 8.5

Move to the Top Right Again: Creating Flexible Conversations About Fairness

So, this brings us back to the top-right quadrant of our group matrix. Now that your group has journeyed around the matrix, you have a shared map of not only some concrete strategies to cultivate fairness in your group, but also of some of the difficult thoughts, feelings, and behaviors that are likely to show up on the left-hand side and get in the way. Your group may now wish to adapt and extend the goals they have in the top right. Usually these goals will be highly specific to the group, as they are in figure 8.5. However, in the remainder of this chapter, we'll briefly present some broad approaches we have seen different groups adopt to increase perceptions of fairness.

IMPROVING COMMUNICATION SKILLS

Equity is about having open conversations about perceptions of contributions and benefits. This takes courage but also skill. When people can speak assertively but not aggressively in the context of the needs of others, they are building equity. Even more critically, fostering deep listening within a group allows people to detect more easily issues of perceived inequity and unfairness and then raise them with the group. Reflecting that someone feels undervalued or exploited can go a long way toward assuaging concerns and building a platform for resolving the inequity.

FOSTER PERSPECTIVE-TAKING SKILLS

What exactly do we mean by "perspective taking"? You can think of perspective taking as having a "cool" cognitive component and a "warm" empathic component. The former is more about understanding how the other thinks and feels, whereas the latter, what sometimes is called empathic concern, is more about feeling the other's pain and wanting to help. There is some evidence that the former may be more helpful than the latter for creating equity in groups. Perspective taking helps people step outside the constraints of their biased frames of reference and reduce egocentric perceptions of what is fair.[23] By contrast, empathic concern can sometimes lead people to favor other individuals in ways that violate principles of equity and justice.[24] And they may ultimately lead to poor group performance. Consider, for example, the leader who gives preferential treatment to one group member for whom they feel sympathy. Although caring for others is an important aspect of prosociality, it must be situated within a broader understanding of what is good for the system. Thus, working on processes that help people better understand the needs, values, and interests of others (such as reflective listening and dialogues to deliberately foster perspective taking) can be helpful.

INCREASING PSYCHOLOGICAL FLEXIBILITY

Helping the group become more psychologically flexible can go a long way toward improving both perceived and actual equity. Being aware of the ways that we automatically assess fairness, and how we can get hooked by our self-serving assumptions and beliefs, allows us to put the discomfort of talking about fairness into context in helpful ways and not let it dictate our behavior.

For example, during the writing of this book, Paul noticed that there were times when he wondered whether he'd be recognized for his efforts. Being able to notice these fears and put them in the context of the shared purpose we all had, as well as of his caring for Steve and David, helped quiet the self-protective voice of his mind. This had real consequences. At one point, David suggested rolling Paul's personal webpage into the broader prosocial.world website. Paul felt as though that might threaten his livelihood and autonomy. But by being able to notice the fear, and then actively seeking to listen to what David was proposing, they were able to find a mutually satisfying solution that both reduced confusion between the websites and provided Paul with the ongoing autonomy to run his website and business. Being able to momentarily tolerate situations that looked as though they were working against his interests in the service of a greater good allowed more prosocial behavior to occur. It is important to note that this was not compromise; rather, it was a conscious shift from reacting to small, immediate issues to consciously responding to larger, longer-term values and goals.

VALUING "CARE LABOR"

Sometimes teams are equitable regarding the formal tasks for which everybody is responsible, but some individuals carry much more of the informal "care labor" than others. Whom do people consult when they're feeling down or angry? Who organizes the birthday parties and the celebrations of new births? All groups require at least some of this sort of discretionary work in order to build a sense of belonging. Often this labor is unrewarded, and often it is completed by women. Taking care to notice and distribute this relational work is particularly important in nonhierarchical groups that have deemphasized the caring role of the manager.[25]

FOCUS ON ROLES AND TASKS MORE THAN ON PERSONALITIES AND INDIVIDUALS

Directing conversations to be more about roles and tasks rather than individuals can foster greater equity. Holacracy, a structured approach to creating more democratic organizations, relies upon a careful and detailed discussion of the *roles* that have authority and accountability rather than the individuals.[26] When purposeful roles are emphasized, it's easier to have a conversation about reasonable expectations of that role, and what benefits should flow to it, than it is to have the same conversation about a person. Focusing on roles can take the self-defensive component out of discussions about equity.[27]

AFTER ACTION REVIEWS

After action reviews, during which people review the effectiveness and equity of projects and initiatives, can help create a climate where people understand that temporary inequities (such as excessive workloads) will be acknowledged and recompensed.

SURVEYS

If the group is highly conflicted over fairness and there are large inequities of power, it can be helpful to conduct an anonymous survey regarding perceptions of fairness in the team. For example, managers often believe they are much fairer than they are. In 360-degree feedback surveys, managers often rate themselves very highly on items like "When managing change, I make extra effort to treat people with dignity and respect," but their staff often don't rate them nearly as highly on the same items.[28] Of course, giving feedback to managers and other staff members on issues of fairness can be highly emotive, so feedback that is more behavioral oriented is more helpful than feedback that categorizes the person ("This is what you did," not "This is who you are").

INVOLVE PEOPLE IN DECISION MAKING

Last, but certainly not least, problems of distributive fairness usually arise from problems of procedural fairness. In the next chapter we'll describe in detail how including those who are affected by decisions in the making of those decisions (core design principle 3) can substantially reduce feelings of inequity. When people can freely and safely contribute to decision making, they are more likely to represent their interests appropriately, less likely to feel exploited, much more likely to be creative and innovative,[29] and much more likely to accept the decision long term even when the outcomes are unfavorable to them personally.[30]

Conclusion

Distributive fairness is so intertwined with issues of trust, decision making, transparency, accountability, and power that almost everything we talk about in this book could be said to be aimed at improving fairness. But that is just the nature of the core design principles. A healthy forest is not a collection of chemicals, climate, and fauna; rather, it's all those things in relationship with one another. Similarly, fairness is not something one can target in isolation from every other group process. Fairness, like health, is an emergent property of a well-functioning system. That said, we decided to keep it as a separate principle for two reasons. First, fairness is so important that we've found it useful to have a placeholder in the Prosocial process for discussing fairness in groups when doing so is really needed. Second, fairness is a much more common issue at the level of groups of groups, where it is easier to fall into competitive, judgmental relationships that seek to exploit others rather than cooperate with them.

In this chapter we reviewed some key distinctions that help shift one's understanding of fairness as a fuzzy, emotion-laden term to a more precise and workable issue within and between groups. We described how, in our experience, distributive fairness is more often a consequence of other issues within groups rather than the primary problem. But when it is the primary presenting problem, it's important to have ways to approach the conversation. Our primary suggestion was not to frame the conversation in terms of fairness but instead in terms of how groups can create a win-win situation that meets everyone's individual needs as much as possible, while also addressing the needs and aims of the collective. When groups have a clear sense of shared purpose, open lines of communication about needs and values, and a win-win culture in which everybody's needs are considered, fairness issues tend to be resolved over time. Key capabilities such as communication, perspective taking, and psychological flexibility underpin the building of trust and help with the more concrete strategies, such as valuing care labor, focusing on roles and tasks more than personalities, reflecting on what has worked, completing surveys, and building more inclusive decision making into a group's process. It is to that last topic that we now turn.

CHAPTER 9

Core Design Principle 3
Fair and Inclusive Decision Making

The Association for Contextual Behavioral Science (ACBS) peer-reviewed acceptance and commitment therapy (ACT) trainers community was formed in 2003 as a means to disseminate findings from the science of behavior change to the world. When the ACBS recognizes trainers, they go on the list. By 2016 the group had grown from a handful to eighty-five, all with outstanding reputations as ACT trainers. But despite this growth, the organization had no formal roles or procedures other than a then twelve-year-old values statement that each new trainer signed. This statement was occasionally revisited, but because no procedures had been established for making decisions (because the group started with only a few trainers, this had been left for future discussion), these revisions were not linked to actual policy decisions.

Tension was beginning to form within the group. There was an academic faction and a practitioner faction, as well as other conflicts about issues such as gender relations. Some trainers felt that others were too aggressive in their commercialization efforts. Overall the group felt like a collection of individuals rather than a team, which in a way was not surprising since the agreements needed to function as a formal group had never been made.

The group found itself locked in a kind of trap. They had no shortage of shared purpose, but no way to execute that purpose. Distributed around the world, connected only by a shared love of ACBS, the work, and small groups of friendships, deciding how to decide together became a daunting task. Matters such as whether or not to issue a formal response to a major mental health concern in the news; agreeing on a shared purpose; or deciding whether to rely on an elected leader, team, or a consensus-oriented process left the group paralyzed. Some members lapsed into inaction due to disagreement and conflict, whereas other disengaged.

The group decided to use the Prosocial process, led by Beate Ebert and Paul, to enhance group collaboration. They began with the Prosocial survey of core design principles and found that they wanted to prioritize a clearer sense of purpose, a way to make decisions together, and a way to manage conflicts between group members. The group formed committees to look at each of these issues. (For the purposes of this chapter, in this example we focus only on what happened in the decision-making committee.)

> *From 2016 to mid-2018, the decision-making committee researched a wide range of decision-making protocols resulting in a specific proposal based upon consent-based decision making. While it would have been much easier to go the usual route of forming an elected representative leadership group to make decisions on behalf of members, the group elected to try a model that increased engagement and inclusiveness.*

If a group is to achieve anything of value together, it must be able to make decisions effectively and efficiently. And if decision making is sometimes hard for individuals, it is much harder for groups of people with widely differing perspectives and interests and who are driven by their emotional systems to defend those perspectives and interests while living in a social climate that's increasingly volatile, uncertain, complex, and ambiguous. Some groups, such as the peer-reviewed ACT trainers just described, seem paralyzed between two suboptimal extremes: centralize power in the hands of a few, or face endless unproductive discussion, deliberation, and uncomfortable disagreement.

Ostrom recognized that groups around the world were more effective when members were more involved in making decisions about the issues that affected them, including the rules for how to work together.[1] When individuals in a commons have a voice that can influence the issues that affect them, they are energized by the act of protecting and advancing their interests (which can, of course, include caring for others and the group). When a group develops the agreements and culture that supports everybody having a voice, groups can thrive as they weave individual interests into collective purpose.

In this chapter, we explore the power and challenges of inclusive decision making (core design principle 3), as well as the forces that hold us back from giving real power to the people. And we explore how we have worked with groups (including the peer-reviewed ACT trainers) to help them avoid the twin perils of excessive centralization of authority on the one hand, and the paralysis of disorganized and inefficient decision making on the other.

The Benefits of Inclusive Decision Making

Inclusive decision making helps both groups and individuals thrive. In a recent review of seventy-two research studies with more than thirty-two thousand participants, Gavin Slemp and colleagues showed that when leaders give followers choices and support them in executing those choices, they are happier, have less distress, are more productive at work, and are more intrinsically motivated to exert discretionary effort.[2] Indeed, studies of the importance of self-determination have shown that it's a basic human need that affects not just work performance but every aspect of life, from parenting to physical health, and from mental health to learning, social development, and cooperative behavior in groups.[3] Let's look in detail at some of the benefits of inclusive decision making.

More developed, skillful group members: Inclusive decision making helps group members develop crucial skills—not just the technical skills of learning about the issues and approaches to solving them, but also the psychological skills of listening, perspective taking, self-regulating, making room for other points of view, and committing to action even when it's scary.

Improved engagement: People who participate in making decisions are more likely to support those decisions. In communities, organizations, and other groups around the world, involving people in change is one of the most reliable predictors of whether the change is successful and sustained. Conversely power hierarchies can have enormous negative effects upon engagement. One study by Katherine Cronin and colleagues placed participants in hierarchical or nonhierarchical conditions, with or without power. Over nine sequential rounds, both high- and low-ranked people acted less prosocially when they were in a hierarchical condition, and lower-ranked participants made fewer and fewer contributions over time.[4] This experiment showed the very essence of self-interest in action. Lower-ranked participants saw that there was little chance of them receiving a fair share of the proceeds, and so they withdrew their effort. Higher-ranked participants then had to work harder and harder to try to reach agreement by making larger and larger contributions, but to no avail. Cooperation simply didn't happen as often in the groups with asymmetrical power as it did in the groups where power was balanced.[5] This result and others like it have been replicated repeatedly in workplace groups.[6]

Better decisions: When a group is cooperative and includes multiple perspectives, better information is used to make decisions. Cooperative groups make better decisions than individuals.[7] People on the front line are more likely to know what is needed, and divergent perspectives often reflect the different risks and opportunities involved in the decision.

Improved cooperation and performance: Sharing responsibility can improve performance by increasing motivation. For example, firms that are owned by their workers outperform their peers,[8] and group members are more likely to support leaders who involve them in decision making and are more likely to stay in groups led by democratic leaders as opposed to autocratic ones.[9]

More resilient systems: Inclusive decision-making systems are more resilient over time because they are less dependent upon individuals and their specific preferences. In a world where everyone's interests matter, we need systems that are resilient enough to handle multiple different perspectives and interests. We need to plan for "difficult" people and have systems that can manage them. Egalitarian teams are also more resilient to interteam conflict.[10]

Improved well-being and vitality: Ultimately, perhaps the most fundamental reason to involve people in making decisions about the issues that affect them is that such self-determination improves well-being.[11] Of course, choosing for oneself can itself be stressful if one lacks information, when all the alternatives have downsides, or when choice involves conflict. But the evidence is clear: people are happier when they have more control over their lives than when they do not. Though choosing can be hard, not being able to choose is much worse.

So, it's clear that inclusive decision making has many benefits to offer. But despite all the blogs, books, and training seminars extolling the benefits of empowering group members, strong hierarchies of authority still persist. Decisions are made at the top and "pushed down" to those who are supposed to execute those decisions. This doesn't just happen in work groups. Despite years of evidence supporting autonomy-supportive styles of parenting over coercive styles, most parents still regularly resort to command and control tactics to influence their children. Coercive behavior only gets worse when those in charge are threatened. Under time or resource pressures, the degree to which they rely upon their authority and power to enforce decisions only gets more intense.

Why is it like this? Surely if more devolved, inclusive decision making is such an obviously good idea, then it ought to be selected for in the competitive world of organizational life, in community groups, even in day-to-day contexts such as parents engaging with the children they love.

If we can understand why hierarchies persist, perhaps we can better understand what we can do to mitigate their negative effects and utilize the benefits of inclusive decision making.

Why Hierarchies Persist

Whenever a behavior persists, we need to look for the consequences maintaining it. Indeed, if we step back and look at the whole system in which a group operates, we can see that there are strong selective forces pushing authority figures to become more dominant and controlling and pushing "followers" to accede to this situation.

First, in some circumstances hierarchies work to create greater cooperation and coordination. Followers of authority figures are usually willing to forgo power in exchange for the direction, protection, and social order created by authority figures.[12] Authority figures are expected to cut through the uncertainty, identify the core issues and problems, and have solutions to those problems. Authority figures are meant to protect their followers from external dangers and provide stability and order within the group. In short, when people give away their power to authority figures, they no longer have to make risky decisions, and they avoid facing risks or conflict.

Second, the self-interest of leaders sustains hierarchies. When people are randomly assigned to leadership roles in experiments, over time they tend to benefit themselves at the expense of others (for example, taking more of a shared resource), often without even being aware of what they're doing. We can see this pattern writ large in the daily press as salary growth among the rich far outstrips inflation or wage growth for the middle and lower echelons of wage earners. Experimental studies demonstrate that such selfish behavior results from an increased sense of entitlement[13] and is exacerbated among those who have more-selfish values systems[14] and in circumstances where an individual in a leadership position within a hierarchy feels threatened or insecure.[15]

Coercive authority is immediately reinforcing in other ways too. Parents yelling at a child to stop annoying them are likely to experience an immediate, short-term reduction in the annoying behavior. Similarly, a boss who tells someone what to do and sees them do it feels more worthwhile and more in control. Human beings love to solve problems, and it is enormously stressful to be out of control, so taking charge and telling people what to do is a way to immediately reduce the discomfort of inaction and uncertainty, while also increasing the positive emotions arising from agency and control.

Third, and relatedly, authority figures are often afraid that they'll lose power if they try more inclusive decision making. We have found from our coaching of leaders that this fear can develop for a variety of complex reasons. Often, leaders, particularly men, who have been socialized to equate leadership with having all of the answers, may feel that involving others in decisions is a sign of weakness and failure. If they believe their value lies in providing solutions, then engaging others in problem solving and decision making means that they will lose their self-perceived sense of value. The same is true in situations where knowledge is perceived to be power—*If I share my understanding, then I lose my power and others gain it.*

Fourth, alternatives to command and control, such as consultation and consensus-based decision making, have downsides that make them unattractive to both leaders and followers. In general, when a larger number of people are included in decision making, the longer a decision can take and the greater the risk of potentially distressing conflict along the way. Even though there is an enormous appetite for creating groups that better integrate individual and collective interests in decision making, many people still seem to believe that the only alternative to top-down decision making is endless, unproductive, and potentially unpleasant discussion.

Finally, power over others can sometimes have more subtle effects that perpetuate command-and-control styles of leadership. One line of research showed that power had effects similar to those of traumatic brain injury! People placed in positions of power became less aware of risks, more impulsive, and less empathic. Other studies have shown that power impairs the neural process of "mirroring," which underpins our ability to understand and empathize with another's perspective. And given that powerful people pay less attention to others, they begin to rely more heavily on stereotypes about others and their own personal vision of what is desirable,[16] thereby fueling the tendency to benefit themselves and adopt a narrower, more self-centric view of the world.

So, there are clearly many factors reinforcing hierarchical structures and their command-and-control operation in the small groups we form. Unfortunately, such hierarchies are well evolved for stable and relatively simple contexts that a single leader can understand, but they do not work well with the complex relationships, distributed information, and rapid change that are the hallmarks of the societies we now live in. Decisions propagated up through hierarchies take time and become increasingly disconnected from the information needed to make good decisions. If we want to create groups that balance individual and collective interests in the long term, we need to do the hard work of empowering people to make decisions and learning new ways of making decisions together.

Integrating More-Inclusive Decision Making into Groups

Inclusive decision making needs to be fit to context. Command-and-control decision making can be helpful in some situations, such as when a group is in an emergency and lacks time or information about how to proceed. When to use this style can also be a question of degree. Not everyone must agree on everything. There's a range of decision-making approaches falling between total autocracy and whole-group consensus. Your challenge is to come up with an approach that involves people in the decisions that they wish to be involved in, while not overwhelming them with decisions they'd prefer others to handle. Shifting this balance of power is an ongoing process of conversation rather than a one-off, one-time change. Let's explore how you can integrate inclusive decision making into your group.

Using the Collective Matrix to Explore Decision Making

Earlier we argued that it's hard to shift decision making in groups. It's hard work because in strong hierarchies, decision makers get to exert power and sometimes benefit themselves while followers get to avoid uncertainty and the responsibilities of decision making. Together this can make for an extremely stable system that is hard to change. Anybody who has worked in hierarchical organizations will know that, while people might *say* it's a good idea to devolve power, it is a lot harder to actually get people *doing* it.

The matrix can prove powerfully useful when there is resistance to creating more inclusive forms of decision making. If a group is to make progress with change, its members must not only talk about what they want to move toward, they must also have a way to talk about the positive and negative reinforcers that are keeping the system stuck. Figure 9.1 highlights the sorts of issues that might arise when exploring this question with a group: What is important to us about fair and inclusive decision making (core design principle 3)?

Core Design Principle 3: Fair and Inclusive Decision Making 141

What can you be seen to DO to avoid or control those difficult experiences?

Get autocratic and shut down discussion

Stop participating—check out during meetings

Rigidly defend my own position

Disengage and not attend

Conflict—overt disagreement

Can you think of any practices that might help us implement this principle?

Small committees that do background research then make proposals

Decision circles

Provide information and consult on the decision

Use a facilitator for tough decisions

Consent-based decision making

Clear roles to delineate who needs to be involved

Communication skills training

Emphasizing and reinforcing shared purpose

Open space and appreciative inquiry

outside ↑

away ← → toward

↓ inside

When a group seeks to implement this principle, what thoughts and feelings show up and get in the way?

From group members

CONCERN that my needs won't be heard or considered

"It will take too long" "It's not my job anyway, the boss should make the decision"

Distrust that my input won't be used

APATHY, DOUBT, RESENTMENT of past treatment

"This is a whitewash, they only care about the numbers" "no one likes my ideas" "I will look foolish if it is a silly idea"

From group leaders

"I will lose control"

"They won't have all the context and then I will have to spend ages explaining. I don't have time!"

Why is this principle important to the group and its members?

Fairness—Everyone deserves a say

Trust

Engagement

Diversity of opinions makes better decisions

Figure 9.1

Models of Decision Making You May Find Helpful

The following recommendations are based upon our observation of international trends in inclusive decision making and our own work exploring decision-making processes in groups that function well. Specifically, we've drawn upon insights from sociocracy, Holacracy, and LaLoux's work on reinventing organizations; resources such as Tim Hartnett's book *Consensus-Oriented Decision-Making*, and Sam Kaner's *Facilitator's Guide to Participatory Decision Making*; and organizations such as Enspiral and Gini that make use of inclusive decision-making practices in organizational contexts.[17]

When designing an approach to decision making, groups need to consider not only who makes the ultimate decision, but who gets to define the decision, issue, challenge, or problem and who gets to provide input on the solution—all factors having to do with the degree of inclusivity. We organized the following options along a spectrum, from least to most inclusive.

Autocratic decision making: The leader frames the decision that needs to be made and announces the decision, perhaps justifying why the decision was made the way it was. While this "tell and sell" approach may sometimes be justified, increasingly it's becoming unviable in modern organizations and many nonwork groups, and it has significant costs in terms of decision quality and engagement.

Consultative leadership: This form has a minimal level of inclusiveness. The leader is skilled enough to consult with, listen to, and understand the perspective of group members affected by the decision. Much of the aforementioned research showing the massive benefits of inclusive decision making is based upon this minimal level of consultation. So even if you or your group operates within a very traditional hierarchical structure, there are things you can do to make a difference.

Facilitative leadership: This is a kind of hybrid we've seen in many traditional hierarchical organizations in which the group leader is skilled in facilitative leadership. The leader not only seeks out multiple perspectives but encourages group members to take real responsibility for finding workable solutions to group challenges. Such an approach might include voting on the decision, and groups might choose to make decisions based on a majority (more than 50 percent agree) or supermajority (higher thresholds for agreement, such as 60 percent or even 90 percent). However, because this all occurs within the context of a traditional hierarchy, the group understands that if it fails to reach an agreement, the group leader is ultimately accountable for the decision and will take responsibility for it.

Up to this point we have been discussing decision making within traditional organizational hierarchies, in which decision proposals usually come from those in charge who also have ultimate authority over the final decision. Important change can occur when members of the group are given real authority to not only make decisions for themselves but to propose entirely new options for the group to decide on. What follows are three variants of this approach that we've seen in action.

The advice process: This approach shifts the power to frame a decision opportunity to members of the group. Anyone in the group is free to make a proposal and then consult with others who will be meaningfully affected, who have expertise in the decision domain, or who have other important information about the decision to be made. You might be thinking that such an approach would lead to chaos, with everyone pursuing their own initiatives, but in practice it is possible to coordinate activities by ensuring that key personnel, potentially including managers and leaders, are consulted throughout the process.[18]

The advice process can seem somewhat autocratic. Proposers are free to decide for their proposal even in the face of strong objections from those they consult. We have seen many examples of leaders using this process to pay lip service to giving real authority to their staff but in practice retaining power over decisions. Such pretense is worse than simply being clear that decision-making power rests with authority figures, and it generally breeds skepticism and resentment.

Consent-based decision making: In some ways, this process is similar to the advice process in that everybody in the group is free to facilitate discussions about an issue and formulate proposals. However, it differs in that authority to decide is distributed across the group. With one model, instantiated in the Loomio online decision tool, as well as those implementing models such as Holacracy and sociocracy, anyone can create proposals after consulting with the group. Then the group votes on the proposal, with these typical response options:

- **Agree:** I agree with this and approve of it moving forward.

- **Abstain:** I am happy for the group to decide without me.

- **Disagree:** I think we can probably do better.

- **Block:** I have strong objections to this proposal and do not approve of it moving forward.

The abstain vote gives people the opportunity to voluntarily decline being involved in decisions that they care less about or on which they are poorly informed. The disagree option gives people an opportunity to express ideas for improvement, without holding up the decision process. The block option gives everybody significant power in the group, thereby avoiding the problems of majority voting that can lead to a significant subgroup being disenfranchised by a decision.[19] However, it will only work in the context of clear guidelines and a positive culture of trust.

Consensus decision making: This form is usually defined by everybody agreeing on the decision. Whereas *consent* means "I can live with this," *consensus* means "This is my preference." As a result, although consensus can work with some small, highly trusting groups, such as families or small tribes, it is usually unworkable in settings with diverse group members under time pressure to decide.

Now, once you've decided on a model for decision making, you now have to work to implement it—to put it into practice. Our experience is that even the best-designed process can go badly awry if the group doesn't have a culture of positive intent, or if group members lack the skills to self-regulate emotionally, understand others' perspectives, or engage in dialogue. Such cases will require adjusting your group's culture. For these groups, psychological flexibility comes in handy.

Culture and the Inner Work of Inclusive Decision Making

Psychological flexibility helps make sharing decision-making power a lot less threatening. Psychological flexibility can help people sit with the discomfort of disagreements and bold decisions,[20] be less attached to defending their positions and identity,[21] and feel empowered to readily step up to make decisions to influence their own circumstances.[22] And when people are more psychologically flexible they're less likely to make impulsive decisions or to act in their own short-term interests at the expense of their long-term goals. Delay is distressing, but learning to tolerate that distress provides one with more space for choice.

Consider, for example, what typically happens in a meeting when people are striving to solve a problem. Someone raises an important challenge and the group starts to brainstorm solutions. For this to be genuinely creative, there must be a divergent discussion phase in which ideas are freely generated and discussed before a convergent narrowing to a decision. Sam Kaner calls the divergent phase the "groan zone," as people start to feel uncomfortable with the uncertainty and start to say things like "We are not making any progress here," and "What is the point of this?" and "Let's just get on with making a decision." It is at this point that many groups give up on inclusive approaches to decision making and either jump to a premature decision or throw the decision back to an authority figure. But with training in psychological flexibility, group members are more able to resist temptation to avoid the discomforts of uncertainty and more likely to remain engaged in decision making that's in line with their values.

Earlier we talked about using the matrix for deciding on the *process* of decision making a group adopts. But it can also be used to discuss the *content* of particular decisions. Simply start with this question: What is important to us about [decision]? Using the matrix ensures that there is repeated opportunity to work on the psychological flexibility of the group and its members. That is particularly critical for the leaders of groups practicing inclusive decision making. Taking responsibility for moving change forward is stressful. Transformational leaders who have psychological flexibility skills are better able to act in the face of difficult emotions, such as uncertainty and doubt, without having to defend their egos, be right, or look good. That frees up attention to listen to others and base decisions on better information and a wider perspective.

Decision Making in More Traditional Hierarchies

So far in this chapter we've explored options for highly inclusive decision making, such as the advice process and consent-based decision making. But many readers of this book will be working with groups in bureaucratic and hierarchical organizations where such processes might seem like a fanciful dream. The whole point of bureaucracy is to control employees so they follow rules, adhere to standards, and meet key performance indicators. This context makes it extremely hard to sustain truly inclusive decision making. But you can make a start. Here are a few options that are easily manageable, even in highly hierarchical organizations.

USE THE MATRIX TO INTEGRATE PERSONAL AND COLLECTIVE GOALS AND VALUES

The matrix is a fantastic tool for clarifying and integrating individual and collective goals and values. When these are clear, people are more empowered to make decisions that bring the two in line. Once you've conducted a collective matrix focused on core design principle 1, shared identity and purpose (see chapter 7), encourage each group member to develop a personal mission, outlining the value they wish to provide to better serve the group's collective goals. Members can share and discuss these personal missions to increase coordination, self-determination, and accountability. Sharing ideas about what gets in the way of these personal missions or role statements (the left side of the individual matrix) can also lead to group activities that make it easier for people to accomplish their missions (for example, by removing sources of fear or mistrust). And when leaders involve themselves in this process, including articulating their commitments and their concerns, this is a fantastic opportunity to decrease the perceived gap between leaders and followers.

IMPROVE DELEGATION PROCESSES

Ultimately, inclusive decision making is less about trying to anticipate what might happen and more about dynamically sensing and responding to changing circumstances. Instead of just telling staff what to do, delegations that outline the broader context for decision making (why this matters to the group and the leader) and any concerns the leader might have both include and honor the person being delegated to. For example, compare the command "Please ring Joe to find out what he's complaining about" with the statement below, which elaborates on the context for this request, the precise task this supervisor wants her employee to perform, her reasons for making the request, and what she intends the request to do:

> Complaints have been increasing week to week [situation]. This week, Joe, one of our best customers, called to complain. Could you please call him to see what happened [task]? My aim is to improve our service [intention], but I am concerned that our new ordering system might be getting in the way [concern]. I think a call might help us improve the situation. What do you think?

Which do you think is likely to create greater motivation and better decision making?

With appropriate attention to the needs of group members making decisions, inclusive decision making can easily become part of daily life, even in hierarchical organizations.

BE AS TRANSPARENT AS POSSIBLE WITH INFORMATION

An extremely effective way to exclude people from decision making is to keep information hidden. Open-book management is a great example of inclusive decision making (https://www.greatgame.com/get-started), with its emphasis on educating staff to understand the impact of their decisions and deciding as if they were responsible for the finances of the group. The Internet has greatly enhanced our capacity to provide information transparently and involve affected people in decision making.

IMAGINE STAKEHOLDER PERSPECTIVES

In situations where it is impossible to include all those affected by decision making (such as when a group member is absent or on leave), it can help to imagine talking about the issue with stakeholders, because doing so invites perspective taking.[23] The self-management consulting firm K2K suggests four key questions to discern if a group is making a good decision, the last of which encourages this kind of imagined perspective taking:[24]

1. Will it be better than before? (Think long term.)

2. Can we explain it to everyone? (Is it transparent or not?)

3. Has everyone who'll be affected by the decision and who has relevant, important knowledge been involved or consulted?

4. If we put ourselves in the shoes of those who'll be impacted, would we make the same decision?

HAVE BETTER MEETINGS

Even though decisions are increasingly made asynchronously through Internet-mediated communication, the meeting is probably still the place where most decision making happens. Meetings in which people listen deeply to one another, take their perspectives, and notice but let go of inner urges to defend themselves or their positions inherently involve people more and are more likely to be useful. In addition, even without changing hierarchical power structures, we have seen groups make meetings much more inclusive and effective by adopting the following strategies.

Changing the reinforcers for speaking up and encouraging "loyal dissent." Talking with groups about what normally happens when they raise a new point of view or take a contrary position in meetings can be extremely informative. In our work with the matrix, people cite many thoughts that seem to get in the way of them contributing, such as "I don't have the ability," "My ideas are not important enough," "I don't know as much as others," "My role in this team is too junior," "It isn't my job," and "Only the bosses have a right to speak here." Some people come from cultures where speaking up for one's own point of view can be seen as disloyal or disrespectful. Some people have a long history of being ignored and have disengaged from the team or, even worse, are actively hostile toward it—"I am angry, and these guys are the last people I would help." Sometimes gender stereotypes prevent one gender from speaking up and the other from listening. And sometimes people are just afraid of the conflict that speaking up might generate— "Things will get out of control." Being rejected activates the same pathways in the brain that "light up" when we experience physical pain.[25] We desperately want to be accepted by others, and we live in fear of rejection. We risk rejection when we speak up, and we are less likely to do it when we perceive that we have lower status in the group.[26]

To encourage inclusivity, we've found it very helpful to simply bring group members' attention to their own ways of being with one another, and then think about the question "How can we make it more rewarding and less challenging for everybody to speak up?" The members of one group Paul worked with talked about the importance of "loyal dissent," which they saw as dissent within a broader framework of cooperation and in the interests of the collective rather than self-interest. Developing this term together took the sting out of conflict and reframed disagreement as a positive rather than a negative.

For some groups, including everybody is less about encouraging speaking up and more about creating a culture in the group that supports that it's okay to be quiet. Being quiet provides space and allows for creativity. If we are constantly filling the space with what we already know, there's no time to consider questions long enough to generate truly new solutions. Forcing people to speak up can create more anxiety, which paradoxically shuts them down.

Managing people who speak too much and dominate. Sometimes people who speak too much are simply unaware of their impact. While exploring the reasons why people didn't speak up with a particular group, we experienced an awkward silence. The reality was that one male group member often reacted with long-winded objections whenever anyone mentioned something new or creative, and group members had given up making such suggestions because of the pain of having to listen to his monologues. Sometimes that awareness can be brought into the meeting with light teasing ("Any ideas? What if we promise Joe won't say it sucks for at least ten minutes? Any ideas now?").

But leaving aside a complete lack of awareness, sometimes people speak a lot because they think it's expected of them, perhaps as a leader or a man. Sometimes people are afraid that if they don't speak they will be ignored, excluded, or misunderstood. Ironically, speaking up too much can come from the same "I am not good enough" space as not speaking up at all. Sometimes issues get bogged down because someone has begun to defend a position, such that their identity is tied up with winning, irrespective of whether it is helpful for the group. If "being right" has been identified as a bottom-left matrix barrier, reminding the group of this can diminish its impact ("Let's all first resolve to let go of being right as we consider this issue so we can share our ideas more openly. Anyone want to start?").

Sometimes groups have found it useful to use "circle talk," perhaps at the beginning of a meeting, allowing everyone the opportunity to reflect on how they are and what is important to them about the meeting, or at the end, so everybody gets a chance to comment on what has happened. Circles, even physical circles of seating, can do a lot to level the playing field in a meeting and encourage everyone's voice.

TALK ABOUT POWER

Inclusive decision making is not just about getting rid of hierarchical control. Every human social interaction includes power dynamics, and it's critical to talk about who has power and how it's expressed in groups if you are to create a truly inclusive climate for decision making. Jo Freeman, in her wonderful article "The Tyranny of Structurelessness," writes about the many ways that removing hierarchical structure can result in more rather than less oppression if active steps aren't taken to include people and balance power.[27] When initiatives to redistribute power simply create a vacuum, and when power has become an undiscussable or dirty word, then people will reorganize into hidden hierarchies of gender, race, or class because that is what they're used to. To ensure that informal power relations don't destroy inclusive decision making, it is important to talk about the delegation of specific authorities and their associated accountabilities, distribution of resources, transparency of information, and everything else associated with power in groups.

Ultimately, Decision Making Is a Question of Commoning

In this chapter our aim wasn't to advocate for any specific model of inclusive decision making. Ultimately, with the Prosocial process we've built all our ideas, whenever possible, around the guiding principle of including people in decisions that affect them. From that core idea you can tack on two other principles that are often helpful:

1. Enable those who have the urgency to take the initiative to make proposals *to actually make them.*

2. Move authority to whomever has the best information, provided they are willing to take the authority.

As we write this chapter, the peer-reviewed ACT trainers are testing consent-based decision making. This is a rich experiment designed to create a greater sense of belonging and involvement for the community. It is entirely possible that individual group members will decide that they don't have sufficient time or energy to be involved in decision making. If enough people come to feel this way, the ACT trainers may wish to shift to a more traditional "representative democracy" approach, such as an elected leadership committee that rotates on an annual basis. Above all else, your group's approach to decision making must be helpful for its context.

This chapter has been all about small groups. But what we've discussed might have much broader implications. Barbara Kellerman, in her book *The End of Leadership*, argues that we can view human history as an unfolding trajectory from top-down to more distributed power: from the time of all-powerful gods, religious leaders, and kings to our present-day distrust of leaders and skepticism about whether they can accomplish anything at all. Others go even further, arguing that even representative democracy is no longer working, and that what's needed is a dramatic increase in decision-making processes that are genuinely participative.

The Internet has dramatically accelerated the extent to which we can share information and make decisions together. And this power can be used to either centralize or decentralize power. We are literally at the beginning of an age when a global commons might be possible, where humanity might choose to create, test, and refine previously unimaginable new models of collective decision making that empower, include, and enable.

We are already seeing large-scale experiments in inclusive decision making. Consider participatory budgeting processes such as those of Porto Alegre in Brazil or Reykjavik in Iceland, where citizens identify spending priorities and determine the budget for neighborhoods, cities, and regions through assemblies and shared decision making. This has led to better schools, renewable energy and recycling programs, and enhanced sanitation.[28] Or the collective engagement of thousands of citizens in designing a restoration plan for New Orleans.[29] Or the massive social experiments of issue-based direct democracy advocates, such as the Flux political movement that aims to use the Internet to replace traditional representative forms of democracy with voting on individual issues that people care about.[30] Of course, as dystopian films remind us, evolution doesn't always proceed to the "good," and we could just as easily end up reacting to our baser instincts to block out the uncertainty and complexity by letting corporations and "great man" leaders tell us what to do. It behooves us all to be creative and flexible, experimenting with new ways to make decisions collectively.

Conclusion

Whereas core design principle 2 is about distributive fairness, core design principle 3 is focused on procedural fairness, specifically, involving in decision making those who are affected by the decisions. We reviewed the numerous advantages of such an approach to decision making, including more developed, skillful group members; improved engagement; better decisions; improved performance; greater resilience; and improved well-being and increased vitality. But despite all these benefits, hierarchies persist—sometimes because they work well but sometimes for reasons that are less helpful for the broader system, such as the self-interest of leaders.

We then reviewed a wide range of options for creating more-inclusive decision making in groups, beginning with the matrix tool, which we used to map what matters to the group and the concerns that members might have regarding inclusive decision making. If we are creative and intelligent, we can come up with better ways to make decisions that include everybody who is affected by the decisions. We ended by exploring how inclusive decision making is part of a much larger global trend that's moving people away from command-and-control forms of leadership toward greater participation in the commons.

In the next chapter, we turn to the critical issue of monitoring agreed behaviors. We'll look at how we can create transparency within and between groups to coordinate action and ensure that the collective interests of the group are best served.

CHAPTER 10

Core Design Principle 4
Monitoring Agreed Behaviors

If a group is to cooperate effectively, members must know what others are doing. If group members don't notice and care about others' behavior, the group cannot appropriately coordinate its actions, discourage disruptive self-interested behavior, or encourage helpful cooperative behavior. We use the term "monitoring" to describe observing with intent to coordinate behavior. For some people, that word sounds a bit sinister because it's sometimes associated with powerful people attempting to control the behavior of others. But when we use "monitoring," we really just mean transparency of behavior—all members being able to see or notice what others are doing in the group. In this chapter, we'll explore what monitoring is and how you can build it into your group in a way that supports individual self-determination while also taking care of the needs of the collective.

Varieties of Monitoring

Monitoring can be as simple as having regular team meetings to chat about what people have been working on, give feedback, and make plans together. Or it can be as complicated as the Joint Operations Center described by General Stanley McChrystal in his book *Team of Teams*.

As commander of the Joint Special Operations Command, McChrystal recognized that traditional hierarchical chains of command were simply too slow to respond to the demands of fighting Al Qaeda in Iraq. Despite deeply entrenched fears of sharing information too widely because of security, he was able to institute a strategy of breaking down literal and figurative walls between different military forces and security agencies in a Joint Operations Center, where different agencies freely shared classified information at multiple levels of the military hierarchy. He writes, "Anyone in the room—regardless of their position in the org charts' silos and tiers—could glance up at the screens and know instantly about major factors affecting our mission at that moment... Any of them, however, could walk freely across the room for quick face-to-face coordination."[1] McChrystal credited this strategy of transparently sharing information about what different groups were doing as a key element in the command's success with Al Qaeda in Iraq, and with creating a "shared consciousness" for the effort.

In our own Prosocial development team, which must interact largely online, we noticed that simply instituting a shared to-do list using online tools such as Notion, Trello, and Slack made our behavior much more transparent. People were instantly better able to coordinate their activities within the team. This allowed fortuitous cooperation to occur as, for example, a team member noticed that synergies could be created between projects, or materials from one project could be reused on another project. But knowing that our work would be visible to others also directly increased our motivation to do high-quality and timely work, or at least to explain what was going on if these aims could not be achieved. Noticing others' behavior improves not only coordination, but also motivation.

Monitoring is central to the success of services such as Airbnb, with its reputational ratings system. By making good and bad behavior visible, it's possible for others to choose how they relate to a person or an entity, and knowing that others will be able to see one's bad behavior is a massive incentive for individuals to behave well. The online discussion platform that we use for our Prosocial community space is called Discourse, and it has a "trust system" built into it that acts as a "natural immune system to defend itself from trolls, bad actors, and spammers."[2]

You can see from our examples that we're not talking here about top-down, controlling forms of monitoring. The word "monitoring" easily invokes images of Big Brother, key performance indicators, and metrics that are implemented coercively, squeezing the meaning and vitality out of work and creating rigid instead of flexible responding. This is *not* what we are talking about in this chapter. While effective monitoring can sometimes be centralized as the responsibility of a leader or a designated group member, addressing it as a collective responsibility, in which all members monitor whether group members are acting in ways that further the collective purpose of the group, and not as a system of top-down control, usually leads to better results. Such monitoring tends to fade into the background and is just a natural consequence of people engaging with each other in cooperative ways. We have in mind here a kind of friendly and supportive monitoring that looks more like taking an interest in someone rather than scrutinizing for compliance. Monitoring from this perspective is just a part of daily life, involving regular contact rather than infrequently "checking up" on people.

The aim of monitoring is to capitalize on the positive reinforcement arising from cooperation rather than the negative processes of coercion and control. Consider fishermen taking turns harvesting an area of the sea, an example from Ostrom's work on common-pool resource groups. An individual fisherman is more motivated and better able to report on whether the previous fishing boat left on time than is a centralized authority that must expend resources monitoring. Monitoring works best when it's built into the efficient coordination of activity.

In other words, this principle is about *transparency*, caring about and noticing the work of others, not introducing a dead hand of control. This is why we emphasize that monitoring ideally include peers knowing what others are doing in the group. In some contexts, this might be as simple as regular meetings to discuss what members of the group are working on. In others, it might involve more formal systems of review, such as after action reviews of team performance, performance appraisal discussions with a manager, or obtaining feedback from a survey of peers and subordinates in something like a 360-degree feedback process.

As with the other principles, this principle works at any scale. A virtual team must be able to see what everyone is doing in order to coordinate and perform, just as global arms control relies upon each nation being able to monitor nuclear testing and distinguish it from earthquakes.

The Benefits of Monitoring

There are many benefits of monitoring.

Increased prosociality: People act more prosocially when they know that others are watching. This is partly a rational cognitive process; obviously, we care about our reputations in the eyes of others and are more willing to cooperate when others are watching what we do.[3] But there are also unconscious processes going on here. In a series of controlled studies, researchers showed that people are more likely to make voluntary donations when there is a picture of a pair of eyes on the wall than when there is not. This basic finding in the lab has been replicated in a variety of contexts: the presence of watching eyes also increases charitable giving, decreases littering, increases the picking up of litter, and increases contributions to an "honesty box" in which people pay for coffee and other supplies in a shared kitchen.[4] A picture obviously cannot tell on our bad behavior, but our psychology is so finely tuned to the presence of others that even a symbol that evokes a sense of being watched can increase prosociality.

Decreased cheating: Children as young as four or five notice and negatively evaluate free-riding behavior in groups.[5] And, even preverbal infants just six to ten months old placed in an experiment with helpful and unhelpful puppet characters can notice and evaluate whether a puppet's behavior is helpful, and then respond more positively to helpful characters.[6]

Increased motivation and shared identity: Of course, prosocial behavior isn't all about just maintaining reputations or avoiding punishment. We like it when others notice and respond to what we do. People hate being excluded and ignored. Having systems in place that allow people to talk about what they're doing builds a sense of belonging and shared identity in groups.

Improved coordination: Even when nobody is deliberately acting selfishly, there are benefits to monitoring, including improved coordination, such as that found in the *Team of Teams* example above. The effectiveness of monitoring depends upon the proximity of group members and their degree of contact. Families, community groups, and work groups rarely collaborate effectively when people are isolated from one another. Consider, for example, the difficulties of coordination and motivation that arise in virtual groups with infrequent contact, or in families in which noticing is restricted by closed doors and screens that dominate awareness, or with children who do most of their behaving in virtual environments with little parental involvement. Transparency enables the coordination of activity, but it also enables social influence to reduce selfish behaviors and increase cooperative behaviors.

How Monitoring Can Go Wrong

Despite its obvious utility, as with all the principles, monitoring is not necessarily something that all groups do well. In our experience, groups do monitoring badly in three main ways:

- To the wrong extent: Too much or too little relative to the purposes of the group

- Ineffectively: When either not enough information or the wrong information is gathered about behavior

- Coercively: When monitoring is used not to enhance mutual learning but instead to force behavior

To the wrong extent: Some groups monitor excessively. Consider some government departments. Departmental heads face very strong punishment both in the media and by their superiors if policy documents and responses contain errors. Under these conditions, some departmental heads implement very tight monitoring of all work that leaves the department. Staff learn that, no matter what they write in their reports, it will be substantially altered "further up the chain" before it's released. Under these conditions, there is little incentive for lower-level staff to create quality reports when they know that most of their work will be cut during the editing process. Furthermore, lower-level staff become demotivated as they see their work tracked too closely, because it sends a strong message that they're not to be trusted.

Other teams don't engage with each other enough, effectively monitoring too little. For example, virtual teams often have difficulty keeping track of the specifics of what team members are doing, so members might end up feeling isolated, ignored, and undervalued. Effective groups either reduce the effort involved in monitoring by making it a natural part of the work (as in, for example, having regular check-in meetings and other efforts at enhancing cooperation), or they increase the benefits derived from noticing what others are doing.

Ineffectively: Ineffective monitoring often happens when individuals within groups don't have the opportunity to see how others are behaving, or when the wrong dimensions of group functioning are selected for monitoring. Academics, for example, are typically rewarded if they publish high-quality papers, get good teaching ratings, and contribute positively to the administrative and outreach aspects of their department. But because an academic's manager may rarely see the academic (except at meetings), performance-appraisal meetings are often a pointless "box-ticking" exercise with the manager relying upon only tangible outcomes, such as the number of papers the academic has published during the year, or their teaching ratings. More intangible behaviors, such as helping colleagues, providing additional assistance to students, and contributing creatively to the growth and development of the department, might go completely unnoticed.

All of the examples we're giving could easily be extended to the groups of groups level. For example, indices such as gross national product are tremendously powerful motivators for the behavior of countries, but increasingly people are realizing that such indices drive behaviors that are not well aligned with what people actually care about, and so they drive problematic behaviors such as excessive consumption and policies that increase inequity.

Coercively: By "coercive monitoring," we simply mean monitoring that's done to force behaviors that are against the interests of the individual. For example, call centers often have extremely tight monitoring of behavior, with measurement of call time to the second being quite common. In some locations, operators are even assessed for the amount of time they spend in the bathroom! You can probably imagine how controlling and demotivating that would feel. The strong message sent by such monitoring regimes is that staff are not trusted to act in the interests of the company unless they are forced to do so. And, of course, such tight and inflexible monitoring regimes often backfire on the organization, with operators responding rigidly rather than helpfully to customers and finding ingenious ways to lower their call time. For example, in one call center Paul worked with, clients with issues that were likely to take more time to resolve were often told that they were being transferred and then put back into the queue for someone else to pick up because helping such clients slowed down the employee's average call-handling rate for the day. Such behaviors work against the interests of the group (in this case, the company) and could be alleviated by approaches to monitoring that engaged rather than coerced staff.

You might have noticed something that these examples of coercive monitoring share. In each instance, the manager is doing the monitoring rather than someone involved in the cooperative work of the group. When someone is placed in a power position in a hierarchy, it's very easy for monitoring to become coercive and for core design principle 3 (fair and inclusive decision making) to be violated. It doesn't have to be this way. Managers can track staff behavior in ways that encourage and support learning and development rather than use coercive means. Organizational hierarchies that reinforce compliance and fail to reward creativity, initiative, and personal responsibility for decision making undermine these positive behaviors.

How to Implement Monitoring Well

There are two key practices to making behavior visible to others in the group. First, groups should share information about what members of the group are doing, either through meetings; online, asynchronous tools, such as project-management software; or other channels. Second, groups need to have processes for managing the performance of all employees that preserve the group's internal cooperation and prosociality.

Meetings

If your group is going to have meetings, it is essential to give some thought to their timing and purpose to avoid meeting simply for the sake of it. Some of the best examples of processes for creating transparency of behavior in groups can be seen in software-development teams using a process known as "agile," which works well for commercial organizations of all kinds that are focused on project work. Within this process, meetings are called "ceremonies" because they involve predefined processes, roles, and responsibilities. It is now common for groups in software organizations to have a series of discrete group meetings using different time frames for different purposes, such as the following examples.

Daily: Meetings focused on checking in that often involve questions like "What did you do yesterday? What are you doing today? What support do you need? What are you doing for your well-being today?"[7] Such meetings can be enormously powerful for creating a context of accountability, learning, and support, and they can be done very quickly (as little as ten minutes) in small groups of people who are so used to the process that they can do them standing up.

Every two to four weeks: Goal setting and reflection meetings in which people commit to action on immediate goals and reflect upon and review what has been learned from previous sprints (a "sprint" in agile terminology is a time-limited project).

Quarterly: Meetings for establishing three to five higher-order, more strategic objectives that link the work of the entire team, coordinate effort, contribute to the group's shared long-term purpose, and determine smaller goals that are then set in daily or weekly meetings. These meetings might also involve reassigning people to teams associated with particular objectives.

Every six to twelve months: Strategic meetings to reflect upon and update the purpose of the group, build more positive relationships through team-building activities, and so on. Such meetings might be held as retreats in an environment designed to support reflection and collaboration.

Groups not focused on project work will have very different needs. A school, for instance, might have a different focus for daily, weekly, term, and annual meetings, but the basic principle holds. When one is seeking to create groups that are purposeful (core design principle 1) and fair (core design principle 2), and that empower people to make decisions about the things that affect them (core design principle 3), it's essential to replace top-down edicts with internal processes, such as meetings, that are structured in a rhythm of engagement and coordination of effort.

Project-Management Tools

An obvious way to make behavior more transparent in a group is to share information about progress made toward objectives using collaborative project-management tools, such as Trello or Asana. A host of similar tools is available. Putting information online in a readily accessible format can be particularly important when members of the group work remotely and the online shared space is sometimes the main means for getting information about what others in the group are doing. This strategy primarily works to enhance coordination, but it can also tap into the other benefits of monitoring, such as by increasing prosociality, decreasing free-riding, and improving motivation.

Processes for Performance Evaluation and Management

Most organizational groups already have processes in place whereby member behavior is reviewed periodically, usually by a manager. Such performance-management processes can work extremely effectively, if the person conducting the review takes a genuine interest in the work of the group member, and if the process leads to real increases in cooperation. Unfortunately, many performance-management processes become rigid and unhelpful, perhaps because the manager has little real understanding of the work of the group member, meetings are too infrequent, or the meetings are treated like cynical box-ticking exercises without any real impacts upon behavior.

For performance management to enhance cooperation and improve performance, it should embody a few key features.

Timeliness: Annual reviews can only sample behavior at the coarsest possible level. More frequent monitoring of behavior is required to provide real-time feedback that is likely to change behavior (see core design principle 5).

A focus on support, not control: In our experience, the most effective examples of monitoring of agreed-upon behaviors were framed in terms of support, learning, and continual improvement. If a meeting feels like an opportunity for mutual learning and improvement, transparency will increase over time. If it feels like surveillance, transparency will decrease.

Mutual: Systems such as 360-degree feedback, in which staff can notice and comment upon manager behavior in the same way that managers notice and comment upon staff behavior, help create norms of mutual accountability and shared continuous improvement. The tendency of managers to act from self-interest, reviewed earlier, is attenuated when followers can give anonymous feedback about their leader's behavior.[8] Making everybody accountable for their behavior both up and down a hierarchy can, in appropriate contexts, dramatically improve the quality of leadership and management.[9] Paul has worked with numerous organizations for which the simple act of introducing bottom-up feedback regarding behavior catalyzed dramatic improvements in cooperation.

Bringing Psychological Flexibility to Monitoring

As with all the principles, monitoring agreed behaviors can be done in a way that is more or less flexible and helpful. To get a sense of what we mean, it's helpful to draw an analogy from monitoring our own behavior as individuals. Earlier we explored how psychological flexibility involves a kind of open, receptive, and curious awareness of one's own experience. Instead of getting angry, depressed, or anxious about one's own behavior, mindfulness-based therapies encourage people to focus more on simply noticing experiences and then acting wisely to change that which can be changed while accepting that which cannot be changed.

This move toward mindful awareness of one's own experience can just as easily be applied to awareness of others. And in groups it's possible to build a culture of open curiosity rather than judgmental awareness of behavior. What seems to help with this is a focus on purpose: How helpful is this behavior for moving us toward what we care about? If it's helpful, what can we do to celebrate and support the behavior, and if it's unhelpful, what can we do to discourage the behavior or learn alternatives that might be more helpful?

In one form of psychotherapy, known as functional analytic psychotherapy (FAP), the therapist pays particular attention to noticing behaviors in session that either move clients toward or away from the sort of person they want to be. So, for example, a client may have come in for treatment for depression. The therapist may notice that a significant maintaining factor for the depression is the highly negative and self-critical inner dialogue in which the client regularly engages. Together with the client, the therapist may identify that excessively self-critical statements are "away" moves, and more self-compassionate statements are moves "toward" the person the client wishes to be. The therapist then has a toolkit for helping the client notice and increase the frequency of toward behaviors and decrease the frequency of away behaviors.

We have seen the same sort of process operating in groups that are highly functional. Instead of coercive monitoring that seeks to judge, criticize, and control, noticing behavior is more functional: "That was helpful. Thank you" and "I experienced that as less helpful, can you try something else?" Accountability can be seen as more of an opportunity for support than for punishment. Instead of "Why didn't you do that thing you said you were going to do?" the noticer can ask, "How can I support you in doing that thing you said you were going to do?"

Make Use of the Collective Matrix

The matrix work also generates information about what needs to be monitored—valued individual and group behaviors as well as defensive individual and group behaviors that are likely to interfere with effective cooperation. For example, a group member might flag in the individual matrix that they tend to go quiet and not volunteer their opinion when they feel unsure about something. This might work against effective cooperation if important dissenting voices are ignored. Given this information, other members of the group are in a better position to notice when the person might be avoiding speaking up and respond helpfully to include that person in the group discussion. Another example we encountered was a leader who noticed that they got more dictatorial and avoidant when they felt insecure. Armed with this information, and a healthy amount of perspective taking and compassion, group members were more able to help the leader notice and improve upon these behaviors when they showed up.

Monitoring agreed behaviors is not just about noticing selfish or unhelpful behavior; it can also help align well-meaning, cooperative efforts that are pulling a group in different directions. Figure 10.1 shows a typical collective matrix that might emerge when group members examine what matters to them about effective monitoring. As with all the other matrix examples we've provided, the words in this one need to be contextualized to a specific group context.

Doing the individual and collective matrices, and learning to pay more attention to their two key distinctions—inner and outer experience and when something feels like a toward or away move—can help sharpen your capacity to monitor your own experience. Practicing mindfully attending to experience, when you simply notice your present-moment experience without judging the experience or yourself, has been shown to improve the capability to more accurately notice what is happening in the present moment.[10] Evidence suggests that deliberately directing attention in this way is likely to increase the degree to which you are able to notice the emotional, physical, and verbal cues being emitted by others and thus be in a better position to respond in helpful ways.

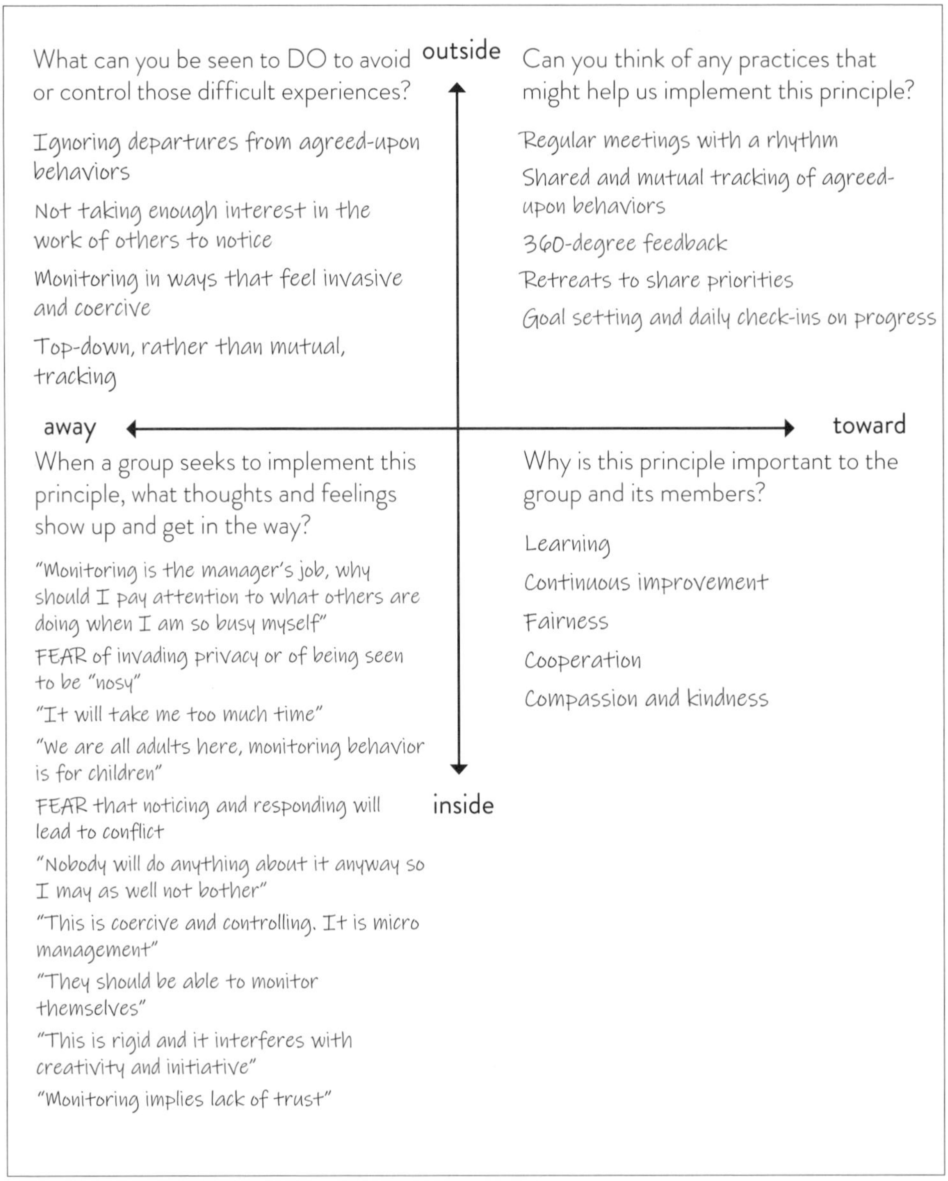

Upper left quadrant (outside / away):

What can you be seen to DO to avoid or control those difficult experiences?

- Ignoring departures from agreed-upon behaviors
- Not taking enough interest in the work of others to notice
- Monitoring in ways that feel invasive and coercive
- Top-down, rather than mutual, tracking

Upper right quadrant (outside / toward):

Can you think of any practices that might help us implement this principle?

- Regular meetings with a rhythm
- Shared and mutual tracking of agreed-upon behaviors
- 360-degree feedback
- Retreats to share priorities
- Goal setting and daily check-ins on progress

Lower left quadrant (inside / away):

When a group seeks to implement this principle, what thoughts and feelings show up and get in the way?

- "Monitoring is the manager's job, why should I pay attention to what others are doing when I am so busy myself"
- FEAR of invading privacy or of being seen to be "nosy"
- "It will take me too much time"
- "We are all adults here, monitoring behavior is for children"
- FEAR that noticing and responding will lead to conflict
- "Nobody will do anything about it anyway so I may as well not bother"
- "This is coercive and controlling. It is micro management"
- "They should be able to monitor themselves"
- "This is rigid and it interferes with creativity and initiative"
- "Monitoring implies lack of trust"

Lower right quadrant (inside / toward):

Why is this principle important to the group and its members?

- Learning
- Continuous improvement
- Fairness
- Cooperation
- Compassion and kindness

Figure 10.1

Conclusion

Being able to see what others are doing in a group increases the likelihood of prosocial behavior. Monitoring also increases the potential for coordinating behavior, thereby increasing the effectiveness of the group. But monitoring can also be done ineffectively or coercively, in which case it can have the opposite effect on the group by diminishing the motivation and performance of individuals. In this chapter we explored different ways that groups can share information about what members of the group are doing through meetings; online, asynchronous tools, such as project-management software; or other channels; and we have explored how monitoring can be done well in support of performance-management processes.

While monitoring is essential for individuals within groups to flexibly respond to one another and act effectively toward the shared purpose, we tried to make it clear in this chapter that monitoring is much more effective when it's used to share information rather than as a tool for exerting power over others. When monitoring becomes coercive, and group members feel a loss of self-determination, there is little hope of creating an effective, cooperative group.

Of course, just monitoring is not much help unless one also responds appropriately to the behavior one notices. If we want to get groups working really well, we need to respond in ways that increase the frequency of behaviors that move the group forward toward its purpose, and decrease the frequency of behaviors that take it away from its purpose. We explore this principle in the next chapter.

CHAPTER 11

Core Design Principle 5

Graduated Responding to Helpful and Unhelpful Behavior

Joan works in a government department devoted to processing claims for unemployment benefits. When she looks around at her team, she can see some people working very hard and others free-riding. There seem to be no consequences, and everyone is treated like a cog in the machine. Over time, she finds herself losing motivation to work hard and instead explores getting by with minimum effort.

Alfred works in a firm selling insurance. His company uses a "rank and yank" approach to performance management. Every year, the performance of all agents is compared against that of others, and the 10 percent with the worst performance are fired. Alfred has learned to do anything to hang on to his job, including falsifying records and manipulating customers to make his figures look good. Everybody he knows feels stressed, and he's not sure he can trust his colleagues who are competing with him, but everybody seems to accept the system as part of the "macho" culture.

Henry volunteers as a coach for a local junior girls' soccer team. He loves coaching and devotes quite a few hours each week to coming up with new and fun training games. Since he took on the role, the team has moved from being at the bottom of the ranking to near the top. The group of parents are generally supportive, but they rarely interact with him, instead dropping their kids off and leaving or spending time on their phones at games. Henry wonders whether he is really making any contribution at all.

These examples illustrate just a few of the many ways that responding to behavior can go wrong in groups. Ostrom's fifth design principle was "graduated sanctions" for violations of rules and agreements. While we recognize that reducing uncooperative behaviors is essential for groups, it is at least as important in most contexts to also pay attention to increasing the frequency of helpful behaviors. Few groups will survive for very long without social support when people act cooperatively.

Core design principle 5, graduated responding to helpful and unhelpful behavior, is a natural extension of the contextual behavioral thinking we introduced in chapters 4 and 5 regarding the ways that behavior is shaped both by the antecedent context and by the consequences that typically followed the behavior in the past. If we wish to enhance cooperation, we need to pay close attention to the consequences we establish for helpful and unhelpful behaviors. Responding effectively to increase prosocial behaviors and decrease antisocial behaviors is not always easy, as seen in how often we get it wrong. It's useful for groups to examine how they typically respond to others' behavior. Are we being fair? Are we avoiding talking about difficult issues? Are we, as a group, taking good performance for granted and neglecting the simple pleasures of celebrating successes and providing support to one another for effort and achievement?

Our key message in this chapter is that the formal and informal processes we put in place to respond to behavior should be focused on creating cooperation to move a group toward its shared purpose, *not* on controlling people. As much as possible, such processes need to honor and preserve the degree to which people feel in control of their choices, their capacity to effect change, and their relationships. We begin by focusing on responding when people act uncooperatively in a group because this was a focus of Ostrom's work.

Creating a Safe Space for Cooperation: Discouraging Uncooperative Behaviors

As we mentioned previously, Ostrom's fifth design principle was "graduated sanctions" for violations of rules and agreements. A *sanction* is an agreed-upon form of punishment for misbehavior. Sanctions can effectively encourage cooperation under certain conditions. For example, we all take for granted the effects of fines and jail terms for discouraging antisocial behavior in society. People are sanctioned for not paying taxes, driving unsafely, or hurting others. Consider what happened in the Ebola crisis in Africa (see the introduction). In some areas, when people didn't cooperate with the new approach to handling the dead, they were first called out on their misbehavior, then they were fined, and then they went to prison. In areas where these sanctions weren't in place, it took much longer for the crisis to pass because people continued with their old habits of washing the dead before burial, and many more people died as a result.

Sanctions can have moderate positive effects on cooperation when it comes to social dilemmas.[1] People are more cooperative when opportunities to punish uncooperative behavior are present than when they're absent, and the threat of sanctions for "bad apples" decreases antisocial behaviors.[2] If you're in a group, it's good to know that "bad" behavior will not be tolerated and may even lead to expulsion from the group; otherwise, group members will lose faith in the capacity of the group to avoid being undermined by self-interest. Furthermore, there's more trust in groups when there are clear and reliable consequences for misbehavior, because people know they are less likely to be exploited.[3]

In Ostrom's formulation, it was critical that the sanctions were *graded* and consistent with the other core design principles. In the common-pool resource groups she studied, initial sanctions for rule violations were often very light. Gossip is an example. When people took more than their fair share others talked about them, and word soon got back to the violators. Similarly, off the northeast coast of the United States, lobster fishermen who place their pots in spaces reserved for others will soon find their floats decorated with frilly pink bows—a way of other fisherman saying, "I see what you are doing, girly man." This message lets those who are breaking the rules know that their violation is unwelcome, that they're being monitored, and that they're being ridiculed.

Sanctions that start small are far less likely to elicit strong emotional responses, counterattacks, or withdrawal from the group. By giving violators an easy way out, the sanctions also are likely to seem fairer and less indiscriminate.

In the successful groups that Ostrom studied, if violations persisted, the sanctions escalated. For example, if gossip did not deter people from taking more than their due, a visit from village elders might follow. To this day, the lobster fisherman who ignores a pink ribbon warning will likely begin to pull up pots that have been destroyed. Sometimes (as in the case of the village elders) increased sanctions will include more guidance about what to do to avoid future sanctions.

If strong sanctions are applied immediately, instead of after a more mild sanction, it can undermine trust and provoke retaliation. Nobody wants to be in a group in which, literally or figuratively, the smallest violation could result in your head being cut off. Graduated sanctions can have the opposite effect, because others see that abusive and costly forms of selfishness will not be tolerated and thus their own cooperative steps are likely to be reciprocated.

The ultimate sanction of being cast out of the group gives gentle warnings part of their punch. We've probably all been in work groups in which we felt a sense of relief or gratitude when a chronically toxic staff member was finally let go.

At the same time, while sanctions can be helpful, they must be treated with care, as poorly designed systems of sanctioning can have serious long-term side effects, such as the following:

- **Counterproductive behaviors:** Threat induces fight-or-flight responding. That is, we may respond aggressively[4] or by withdrawing and disengaging[5] if the sanction is focused on the person rather than the behavior, or if the sanction seems unfair, excessive, inconsistent, or indiscriminate.[6] Furthermore, because sanctions are focused on what the person should not do rather than on what they should do, they may lead to undesired behaviors, such as a person gaming the system to avoid punishment while still not engaging in helpful behaviors.

- **Diminished well-being and engagement:** Too much emphasis on sanctions increases the possibility that people will change their behavior out of fear and avoidance, negatively impacting well-being, engagement, and creativity.

- **Diminished trust:** If deployed in a way that undermines a sense of shared intent, sanctions can also undermine trust in groups in the longer term. The stronger the sanctions, the more detrimental these effects.[7]

- **Inefficiency:** The negative effects of sanctions often mean they require continuous supervision, and the transgressor is likely to return to the unhelpful behavior when nobody is watching.

- **Negative effects on the one administering sanctions:** Someone has to exert effort and time to monitor behavior and administer sanctions. In some instances, this can be highly costly in terms of time or emotions for the person doing the monitoring. Few people enjoy forcing people to comply through punishment. Witness the negative and traumatic effects upon policemen and policewomen of lifetimes spent enforcing the law. There aren't negative effects associated with positive rewards.

Given these challenges, a generalized version of core design principle 5 requires that groups provide positive reinforcement for helpful behavior to both recognize discretionary contributions to the group's shared purpose and to encourage people to remain in the group. When it comes to responding to behaviors in a group, it's important to also encourage "toward" behaviors wherever possible rather than just focusing on the behaviors that are not wanted.

Building Engagement and Motivation: Encouraging Cooperative Behaviors

When a group responds effectively to encourage cooperative behaviors, motivation, goal achievement, and engagement within the group are increased. As you might expect, there has been extensive research in this area, and there are good guidelines for how to effectively offer encouragement. Interestingly, our intuitions regarding what is helpful may not always be correct.

Consider a new group leader who institutes a reward system to encourage cooperative behavior. The "best" performer in the group for any given month is acknowledged in a meeting at the end of the month and given a pair of movie tickets. Will this improve cooperative behavior within the group?

In some circumstances it might, at least in the short term. At least the helpful behavior has been noticed and winners get to enjoy a free movie. Crudely done, however, approaches to rewarding helpful behavior that are coercive, reward selfishness, or ignore the other core design principles can have pernicious effects.

For example, attempts to encourage helpful behavior can go wrong if they undermine the recipient's sense of self-determination, competence, or belonging. Motivation researchers Richard Ryan and Ed Deci have convincingly demonstrated this effect many times.[8] They emphasize how people often associate extrinsic rewards (rewards that aren't integral to the task itself) with someone else seeking to control their behavior. Any system of reward that is not fully transparent (core design principle 4, monitoring agreed behaviors) and fair (core design principle 2, equitable distribution of contributions and benefits), or is wielded by an authority without a sense of consent and control (core design principles 3 and 7, fair and inclusive decision making and authority to self-govern, respectively), can easily lead to feelings of manipulation and control, undermining positive effects. Furthermore, we all seek to explain our own behavior. If we come to believe that we are acting purely for the reward, rather than because we want to cooperate or to work together to achieve an outcome with which we identify, we're less likely to cooperate in the first place and much less likely to cooperate when the reward is removed.[9]

So, how can we establish healthy and helpful patterns of responding to encourage cooperative behaviors in groups? There are several guidelines we can apply.

Encourage positive social interactions: These might include creating opportunities to notice and appreciate the work of others, celebrate successes, or say thank you for a job well done, and even kind supportive words when things go wrong despite the group member's best efforts. As long as these interactions are distributed fairly (core design principle 2)[10] and are linked to behaviors that foster shared purpose and identity (core design principle 1) and self-determination (core design principle 3), they can be powerful aids for fostering cooperation.

Meet individual needs and interests: When the personal needs, values, and interests of members of the group are well understood by others, it is easier for those others to respond appropriately to support cooperation. If, for example, everyone in the group knows that James doesn't really value praise, but he does care about learning opportunities, or if they know that he has a hard time controlling his anger even though he is sincerely trying to do so, it's easier to understand James's behavior and respond appropriately to increase the likelihood of James behaving cooperatively.

Avoid destructive within-group competition: Within-group competition can be destructive if the gains that result from increased winner motivation are canceled out by losses in motivation on the part of losers. It's also destructive if increased competition undermines trust and communication as incentives shift from overall group effectiveness to one individual "winning" over others. Competition can drive group members to focus on individual rewards, diminishing the effort they put into working toward a shared purpose (core design principle 1). This can lead to people wasting time advertising their achievements or, worse, hiding their mistakes and failures, thereby further diminishing honesty, trust, and shared learning within the group. Focusing responding on behaviors and achievements that support shared purpose and identity, equity, and inclusiveness bolsters the group against these dangers.[11]

As we've explored so far in this chapter, discouraging uncooperative behavior and encouraging cooperative behavior is not as simple as sanctioning bad behavior and providing rewards for good behavior. Responding that feels coercive or inequitable, or that encourages selfishness, might backfire and produce unintended consequences. This is an area of group functioning where the power of integrating Ostrom's principles rather than treating them separately becomes clear. Sanctions and reinforcement work more effectively when all the other within-group core design principles (1 through 6) are in place, because they keep the focus on the steps that lead to success of the group as a whole. Conversely, when groups overlook the work that needs to be done with some of the core design principles, both sanctions and rewards pose greater risks to group effectiveness. In the next section, we explore how responding effectively to behavior involves all of the principles.

Responding in Line with the Core Design Principles

If there's a clear sense of shared purpose and there's strong group identity around that shared purpose, then it is possible to make it clear that sanctions are only justified when an individual's behavior works against the shared purpose. Members who buy into the group's shared purpose are more likely to also buy into the need for appropriate responding to discourage behaviors that work against it. The same is true for encouraging cooperative behavior. Responding is more likely to be effective when it's consistent with purpose (core design principle 1), such as heartfelt praise for a job well done, support for skill improvement, or new opportunities to expand or vary one's role in the group.

Paul once worked with a group of call-center operators who were very closely monitored. Every call they answered was measured to the second, and they received feedback five to ten times a day on their average call-handling time. As a consequence of this very close monitoring, operators hung up on customers whose problems would take too long to solve, or they transferred them to another operator. When Paul was able to reconnect them with their purpose in the work (serving customers), the operators recognized how they felt badly about treating customers this way. This discussion resulted in the group finding new ways to track call times, as well as have friendly rivalry, neither of which fed the unhelpful behaviors that harmed the goals and values of the group as a whole. In other words, by changing the meaning of their work they changed the payoff structure for more cooperative behavior.

It is also important to ensure that sanctions are fair and proportional (core design principle 2). Ostrom emphasized that graduated sanctions must be proportional to the seriousness of, and the reasons for, the transgression. For example, consider how a manager might respond to a team member who made a mistake. Jumping right into disciplinary action is likely to shame, embarrass, and demotivate the team member, whereas starting with a simple and compassionate question, such as "Hey, I noticed you did X. What was going on there?" or "Are there ways we could work together to help minimize mistakes in the future?" might more effectively diminish the likelihood that the behavior is repeated, while keeping the prosperity of the group itself central to the conversation. Of course, at the other end of the spectrum, it must also be clear to group members that persistent and serious infractions of agreed-upon behaviors will be dealt with more severely, including the potential for exclusion from the group. Having no consequences for serious misdemeanors cultivates a climate of disregard for cooperation, but having excessively serious consequences for honest mistakes undermines trust and cooperation.

It's also important that those in power aren't solely responsible for determining and administering sanctions. Instead, everyone should be encouraged to take responsibility for sanctioning misbehavior (core design principle 3, fair and inclusive decision making). Sanctions are often more effective if fellow group members administer them rather than a centralized authority. Fairness and inclusiveness breed trust, and sanctions work more effectively when there's trust. For example, in high-trust societies such as Denmark and China, the prospect of punishment for failing to contribute to shared public goods promotes cooperation more effectively than in low-trust societies.[12]

The positive effects of rewards are enhanced, and their negative effects are less likely to occur, when rewards are equitably distributed (core design principle 2), and when the group supports its members' individual autonomy by including them in the decisions about how to allocate rewards (core design principle 3). This is true, in part, because these steps reduce the likelihood that members will feel like they're being manipulated or controlled, and, in part, because this approach better manages the dialectic between individual and group interest. When helpful behavior is regularly noticed (core design principle 4) and responded to (core design principle 5), it becomes part of the culture of the group as a whole, and the social environment itself becomes more trust based.

Sanctions are much more effective in groups that have effective approaches to conflict management (core design principle 6) in place. Responding to unhelpful behavior is often emotionally challenging. In our work with groups, we've found that fear of conflict often drives either excessively harsh or excessively avoidant responses to unhelpful behaviors. Having good skills and social practices in place to manage conflict is critical for a group to effectively respond to unhelpful behavior.

Finally, psychological flexibility supports all the core design principles. When we can absorb the emotional challenges of growth and direct attention toward how to create values-based habits, group members are more likely to engage in all of the principles, and this synergistic effect can become a self-amplifying pattern of evolutionary development.

How to Do It: Creating Effective Responding

Though we've offered some general advice about responding (emphasize what you do want over what you don't want whenever possible, focus on building the intrinsic rewards of cooperation, devote effort to social reinforcement, build responding that is consistent with the other principles, and so on), our aim here is not to propose exactly how your group should respond to helpful and unhelpful behaviors. Instead, we see the collective matrix as the perfect tool for your group to develop its own, contextually appropriate ways of responding to behavior within the group.

One of the many benefits of the matrix is that it reveals what is intrinsically rewarding to people. A more technical way to say this is that this work reveals what members of the group will consider reinforcing. Everybody is different, and we can never assume that a particular behavior will be reinforcing. Paul once talked with a teenager who had just been "rewarded" with his name and photo on a plaque at the local McDonald's, yet he was mortified that his friends might see it! Similarly, educational opportunities, a promotion, or a trip overseas to a conference might all be more of a burden than a form of support for some people.

In practice, the experience of responding to helpful behavior is very different from responding to unhelpful behavior. Whereas the former is usually (although not always) pleasurable, the latter can often feel deeply uncomfortable and raises quite different fears and concerns. So let's look at the sorts of issues that groups raise about responding to behavior when they explore core design principle 5 (graduated responding to helpful and unhelpful behavior) using the collective matrix. We prefer to revise slightly the order we go through the matrix when we explore this principle:

1. What is important to us about this principle? (bottom right)

2. How might we implement this principle? (top right)

3. If we were to do these things, what hooks might show up and get in the way of us fully implementing the principle? (bottom left)

4. And what might we do when these hooks drive our behavior? (top left)

5. Can we refine what we do in light of these hooks and defensive behaviors? (back to the top right)

6. And so on, refining and deepening the collective matrix until there is a clear sense of priorities for action.

We often use the matrix to explore responding to helpful and unhelpful behavior separately, because the values and concerns are often quite different. Let's first look at some of the typical responses we get when discussing responding to unhelpful behaviors. Starting with the bottom right, group members often value this principle if it supports values and overarching goals, such as personal accountability, respect, and compassion. Often groups know what works for responding to unhelpful behavior in a graduated way, such as by first asking the person why they did what they did, followed by a clear escalation process (see the next chapter), and perhaps skills training in responding constructively. The typical issues that stop people from responding constructively to unhelpful behavior include a sense that they don't know what to do, that the response might escalate conflict, that responding is not the person's role, or the plain old fear of confrontation and social disapproval. When these sorts of hooks go unexamined, it's easy for groups to avoid saying anything about a particular behavior, to gossip, or to overreact with excessively harsh and judgmental responding (see figure 11.1).

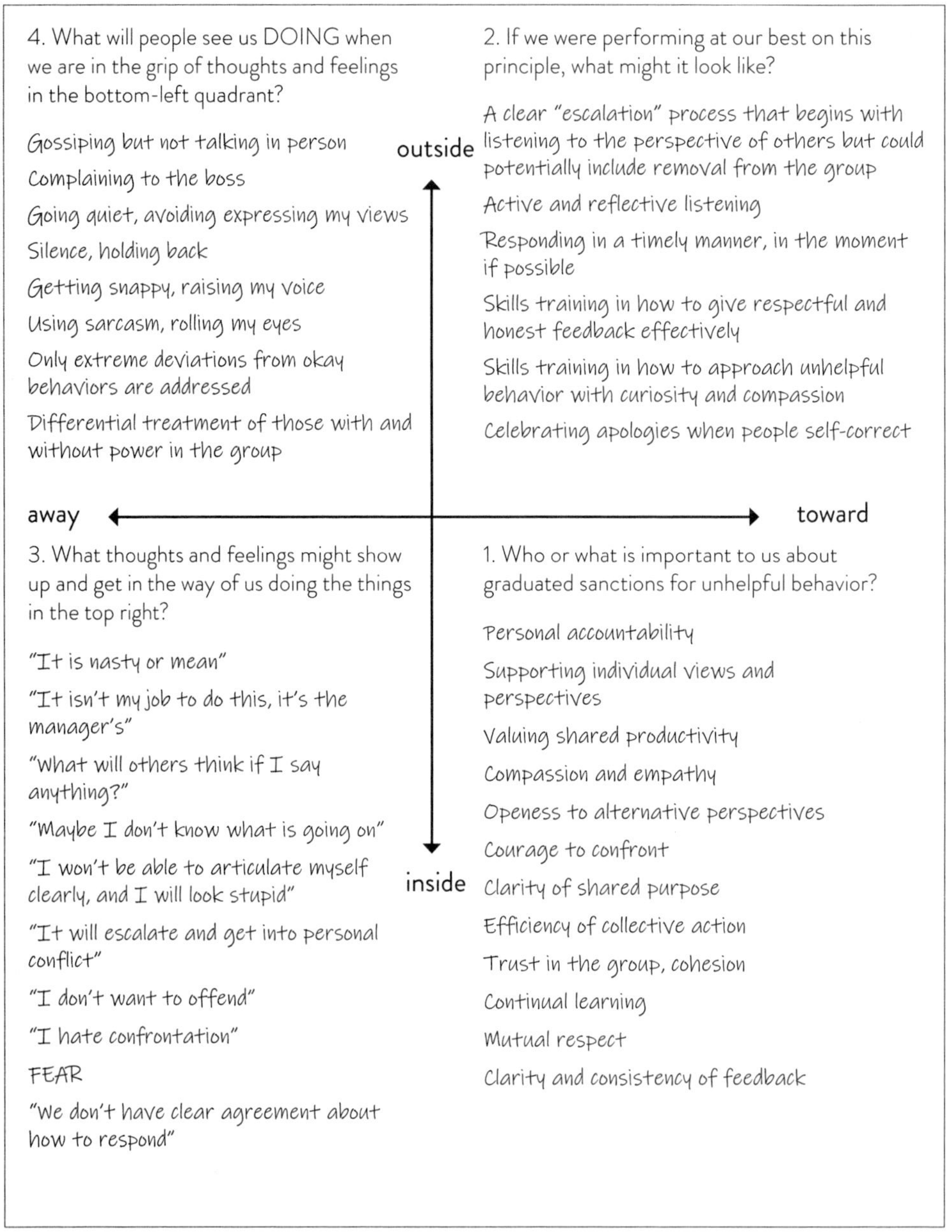

4. What will people see us DOING when we are in the grip of thoughts and feelings in the bottom-left quadrant? Gossiping but not talking in person Complaining to the boss Going quiet, avoiding expressing my views Silence, holding back Getting snappy, raising my voice Using sarcasm, rolling my eyes Only extreme deviations from okay behaviors are addressed Differential treatment of those with and without power in the group	**2. If we were performing at our best on this principle, what might it look like?** A clear "escalation" process that begins with listening to the perspective of others but could potentially include removal from the group Active and reflective listening Responding in a timely manner, in the moment if possible Skills training in how to give respectful and honest feedback effectively Skills training in how to approach unhelpful behavior with curiosity and compassion Celebrating apologies when people self-correct
3. What thoughts and feelings might show up and get in the way of us doing the things in the top right? "It is nasty or mean" "It isn't my job to do this, it's the manager's" "What will others think if I say anything?" "Maybe I don't know what is going on" "I won't be able to articulate myself clearly, and I will look stupid" "It will escalate and get into personal conflict" "I don't want to offend" "I hate confrontation" FEAR "We don't have clear agreement about how to respond"	**1. Who or what is important to us about graduated sanctions for unhelpful behavior?** Personal accountability Supporting individual views and perspectives Valuing shared productivity Compassion and empathy Openess to alternative perspectives Courage to confront Clarity of shared purpose Efficiency of collective action Trust in the group, cohesion Continual learning Mutual respect Clarity and consistency of feedback

Axes: outside (top) / inside (bottom); away (left) / toward (right)

Figure 11.1. A typical collective matrix exploring the issue of responding to uncooperative behavior.

Figure 11.2 illustrates how one group responded when we explored responding to cooperative behaviors with them. When exploring this topic using the matrix, groups typically generate many of the same values, such as mutual respect, caring for one another, and working together effectively to move toward a shared purpose, but they also generate new ones, such as celebrating successes and building engagement and creativity. Some of the specific practices may overlap too. Constructive feedback may remain important to the group, but talking with people about what is most important to them, identifying strengths, and having meetings to celebrate success might be more closely linked with encouraging helpful behavior. Because responding to positive behavior is usually less challenging than responding to unhelpful behavior, the hooks that show up are usually different. Group members often raise issues such as believing that it's insincere to congratulate someone, or that acknowledgement isn't required when a person is simply doing something that's part of their job description, or that responding positively will take too long and detract from the "main work" of the group. When these hooks grab hold, group members don't bother to give positive feedback or take the time to respond favorably to helpful behavior.

The values-based approach we advocate for throughout this book is a great way to discover how best to encourage helpful behaviors. Having worked through the possibilities for positive and negative responding, people in the group will know what works well with other members. It might be acknowledging someone in a meeting or just giving them a pat on the back in conversation. It might be giving them access to new learning and development opportunities, a promotion, or a pay raise—whatever the person values. And when it comes to determining what this is, having done the individual matrix work will pay off again. To the extent that each member of the group knows what other group members care about, they are in a better position to provide reinforcement.

We mentioned earlier that responding to uncooperative behavior (such as underperformance or transgressions) can be particularly challenging because sometimes it involves strong judgments and emotions. In our experience, groups benefit from adopting a staged approach to responding to such behavior, as we describe in the remainder of this chapter.

Step 1: Reorienting to Purpose

When choosing a response to any behavior, it's helpful to orient yourself with this question: What is most helpful to move us toward our common purpose? Often this is a question of learning: What can we learn from this situation to change it in the future? There is sometimes a tension between individual compassion and group equity in responding. Consider the following situation.[13] Imagine the performance of a group's organizational member falters because her father becomes very ill and dies. Though her performance for the year's first three quarters was very high, the fourth quarter reflects a significant drop in rating. When it comes to an "employee of the year" evaluation, an organization that uses all four quarters as a metric would clearly drop her from consideration, whereas those that promote and expect compassionate behavior might find it unfair to exclude her from the running.

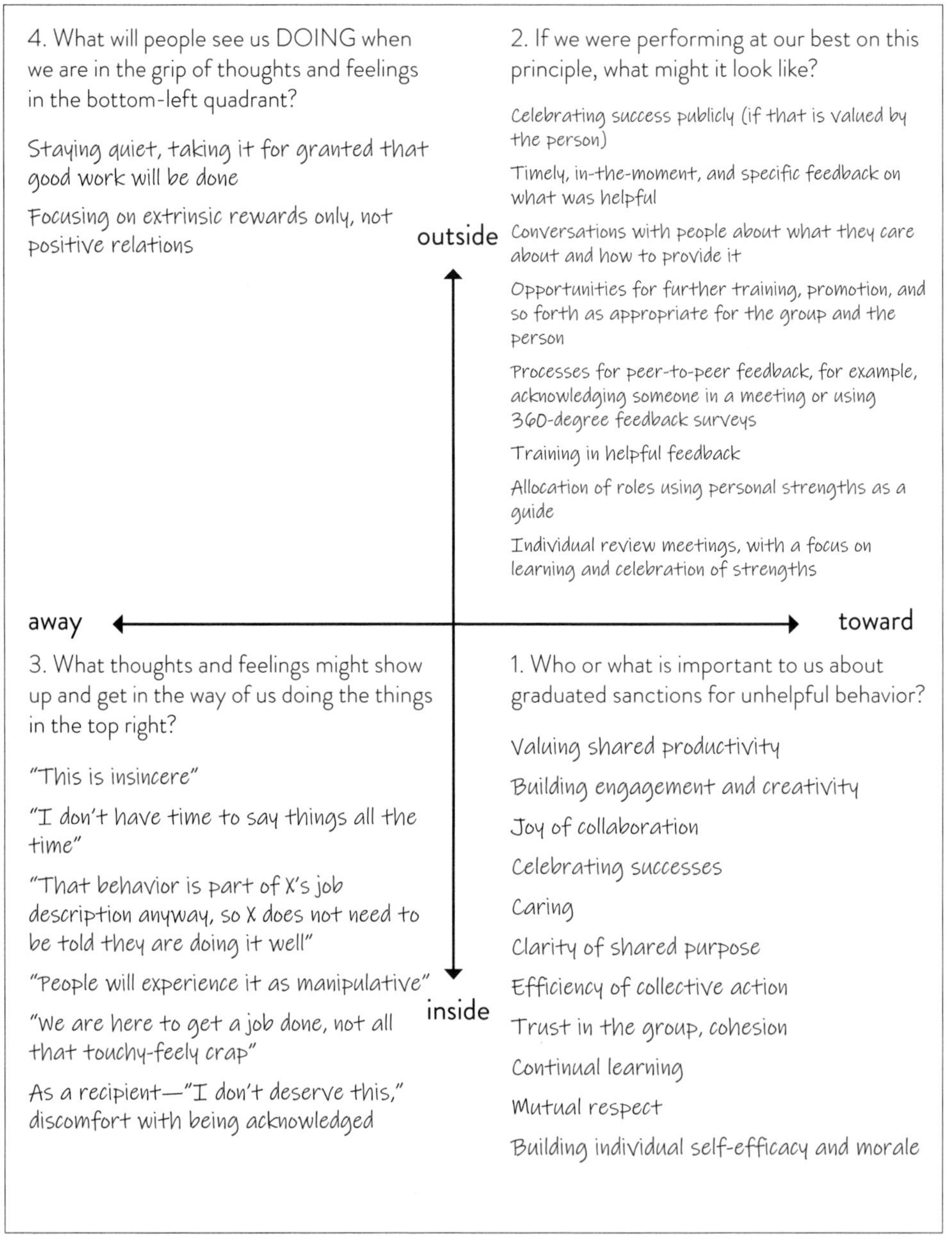

4. What will people see us DOING when we are in the grip of thoughts and feelings in the bottom-left quadrant? Staying quiet, taking it for granted that good work will be done Focusing on extrinsic rewards only, not positive relations	2. If we were performing at our best on this principle, what might it look like? Celebrating success publicly (if that is valued by the person) Timely, in-the-moment, and specific feedback on what was helpful Conversations with people about what they care about and how to provide it Opportunities for further training, promotion, and so forth as appropriate for the group and the person Processes for peer-to-peer feedback, for example, acknowledging someone in a meeting or using 360-degree feedback surveys Training in helpful feedback Allocation of roles using personal strengths as a guide Individual review meetings, with a focus on learning and celebration of strengths
away ←	→ **toward**
3. What thoughts and feelings might show up and get in the way of us doing the things in the top right? "This is insincere" "I don't have time to say things all the time" "That behavior is part of X's job description anyway, so X does not need to be told they are doing it well" "People will experience it as manipulative" "We are here to get a job done, not all that touchy-feely crap" As a recipient—"I don't deserve this," discomfort with being acknowledged	1. Who or what is important to us about graduated sanctions for unhelpful behavior? Valuing shared productivity Building engagement and creativity Joy of collaboration Celebrating successes Caring Clarity of shared purpose Efficiency of collective action Trust in the group, cohesion Continual learning Mutual respect Building individual self-efficacy and morale

Vertical axis: outside (up) / inside (down)

Figure 11.2. A typical collective matrix focused on responding to cooperative behavior.

What would you do in this situation? The answer will, of course, depend upon the outcomes you're seeking and your purpose as a group. Do you want to emphasize a standardized set of rules that people can rely upon to be applied universally, or do you want to be more responsive to individual circumstances, perhaps risking perceptions of unfairness? Such complications are why there can never be one right way to respond to behavior—it always depends on the shared aims of the group.

Step 2: Getting Prepared Personally

When someone does something we perceive as unhelpful, we're likely to have an emotional reaction. Becoming aware of our own reactivity, and using some of the mindfulness and psychological flexibility techniques mentioned earlier in this book, can help you defuse from the emotional reaction and develop a more open and receptive stance for listening to the perspective of the transgressor.

This is not always easy to do. When we believe that a person is deserving of punishment, it's hard to be compassionate. And sometimes we can react harshly simply out of fear. We know that the person is not going to react well, so we act in controlling and punitive ways that are designed to minimize any kind of responding that will expose us to pain. The first step in responding is defusing from judgments about what happened and creating openness to gathering new data. It is important to separate out evaluations that are subjective from actual observations. It's usually possible to separate these by asking, "Would everybody agree on this as a fact?" Calling someone lazy is an evaluation, but reporting that they had shown up late to work three times in a week is a factual observation. Being able to separate fact from evaluation leads us to our next step, finding out what happened from the other's perspective.

Step 3: Stepping into the Shoes of the Other

Unfortunately, as responders to unhelpful behavior, we can easily slip into thinking that the problem rests entirely with the other, not with ourselves. We sometimes think that our judgments about why others did what they did are the truth and, since we already know we are right and the other is wrong, we don't need to engage in the messy and sometimes uncomfortable process of actually talking to them about how they saw the situation or what motivated the behavior.

One of the most robust findings in social psychology is that we tend to attribute bad behavior in others to personal intent, when in fact it is often due to situational constraints. For example, we might assume that a person was late because they were lazy, not because the bus they normally caught was late. In other words, we assume they were acting out of self-interest when in fact they weren't.

For many years Paul used a short role-play drawn from the work of Barry Jentz called the "Frank case."[14] In the case study, participants were asked to play the role of a team leader giving feedback to a team member (Frank) who had interrupted them twice and joked three times during a meeting. Paul played the role of Frank, and the people in his classes role-played giving him feedback. The classic management advice for this sort of situation is to simply describe the situation as factually as possible, describe how it affected you, and then ask for the other's perspective (for example, "Frank, during the meeting you interrupted me twice and made jokes three times. I felt frustrated that I did not get through the agenda. What was going on?"). But *almost nobody* spontaneously approached the situation in this open, learning-oriented way.

Almost everybody assumed they were in the right and Frank was in the wrong. Sometimes they did this aggressively (for example, "Frank, I was very disappointed that you interrupted me throughout the meeting. I do not want you to do that again…"). And sometimes they did it avoidantly, hoping all the while that Frank would spontaneously recognize and repent for his transgression (for example, "Frank, how do you think the meeting went? Do you think anything was improved?"). But almost nobody seemed interested in finding out *why* Frank did what he did. In the case study, it turns out that Frank was trying to help the group leader by lightening the meeting with some humor because he'd heard from others that they felt their meetings were a bit dull. But almost every participant missed this point because they never sought to actually listen to Frank's perspective. Had that question been asked, a totally different conversation might have ensued, and Frank's interest in helping could have been both acknowledged and channeled into more effective forms of cooperation.

Understanding the value of empathy and perspective taking in responding to unhelpful behavior is key. Even the most extreme case of being exiled from a group (that is, being dismissed from an organization) can be done in a way that is harsh and punitive, or it can be done in a way that maximizes opportunities for the person leaving the group to learn and grow from the experience.

Step 4: Choosing a Response

If there's a need to administer a sanction, it's often helpful to simultaneously provide opportunities for reinforcement. In other words, it's always more helpful to accompany a "don't do this" with an alternative behavior that would more effective or helpful.

Second, and in line with core design principle 3 (fair and inclusive decision making), it's always helpful to involve the person concerned in the design of an appropriate response that would in fact decrease the likelihood of such unhelpful behaviors in the future. That isn't always possible, of course.

Once again, the individual matrix can be used to help coach the individual to explore what mattered to them in the situation and some of their internal experiences that led them toward what they did. This strategy is likely to help create optimal conditions for learning and for future behavior change.

Step 5: Implementing the Response

It is important that there's follow-through on the agreed-upon response over time. This creates trust among all concerned in the group that transgressions will be responded to in a timely and effective manner.

Conclusion

In this chapter we provided a very broad review of what's important when responding to both helpful and unhelpful behaviors in groups. We emphasized the need in many voluntary groups to focus on positive consequences, not just negative ones, to build belonging and motivation. Sanctions, when used, need to be graduated and integrated with the other core design principles and flexibility skills. The matrix can be used to help manage these processes. If you'd like to learn more about responding effectively to behavior in groups, we recommend two useful resource books: *Performance Management: Changing Behavior That Drives Organizational Effectiveness*, and *Unlock Behavior, Unleash Profits*.[15]

CHAPTER 12

Core Design Principle 6
Fast and Fair Conflict Resolution

In the poem "Outwitted," Edwin Markham wrote the following lines:

> He drew a circle that shut me out—
> Heretic, rebel, a thing to flout.
> But Love and I had the wit to win:
> We drew a circle that took him in!

Markham was pointing to the way that we exclude others as being wrong or heretical, and that it takes a conscious act of willingness to draw a circle to take them in, listen, and respect and recognize their perspective as a possible perspective.

You might recall the pharmacy case study we presented in chapter 5. Paul had been called in to resolve the bigger conflict that was tearing the pharmacy apart. Kai (one of the lead pharmacists) and Lateefah (another lead pharmacist) had stopped talking directly with each other. Kai believed Lateefah was "bossy" and "aggressive," while Lateefah believed Kai "felt entitled" and was avoiding the issue, "as Asians do." Janine (the CEO) believed Kai was a "good employee" but was "faking" an injury that had led her to make a claim for workers' compensation, and that Lateefah was her "potential replacement" but "not yet very mature." And so on and so on, expanding out to all eighteen employees in the pharmacy. As is so often the case, while the conflict might have started with disagreements about how to do specific tasks, it had long ago escalated into a significant relational issue with each person having fixed beliefs and rigid evaluations about personality, identity, and motivations, and everybody wanting everybody to *know* that the fault lay with the *other*.

Conflict is normal and inevitable in any high-performing group. Group members perceive situations differently—noticing different "facts" and putting them together in different, sometimes self-serving ways. And, as we have stressed throughout this book, people have different interests, values, and expectations about what can and should happen. In Ostrom's work around the globe she recognized the universality of conflict, and so she included fast and fair conflict resolution as a core design principle in its own right. Some authors prefer the term "conflict management" over "conflict resolution," recognizing that some conflicts are intractable and that, even if a particular conflict is resolved, new conflicts will emerge. We have chosen to use the term "conflict resolution" because the processes we use in groups are, in fact, designed to resolve conflicts whenever possible. We added the qualifiers "fast" and "fair" because they are key components of effective conflict resolution.

The contextual behavioral theory of behavior change that we presented in chapter 5 makes it clear that this network of beliefs and evaluations is simply our language machine in action. We create and derive networks of meaning to understand and predict the behaviors of others. But our networks of meaning can easily become rigid when we see them as the only possible truth rather than a perspective ("My perception of X isn't just how I am making sense of the world. It is the way she truly is!"). Furthermore, the sense we make of situations tend to be self-serving—we seek out and process information in ways that will confirm our existing views. Conflict is maintained in our heads, in the judgments of right and wrong, good and bad, and justified and unjustified, and stepping into the perspective of another, seeing things from their point of view and drawing a circle that takes them in, is one of the hardest things in the world to do.

If a group is to find a way through conflicts, it needs to ensure it has effective capability at three levels:

1. Interpersonal skills such as listening well and speaking assertively, not aggressively

2. Personal skills such as emotion regulation and perspective taking

3. Group-level agreements and processes for managing conflict efficiently and effectively

In this chapter, we will consider each of these three practical pathways for ensuring conflict is handled well. We will once again use the collective matrix as a way of organizing our thinking and to illustrate its use for finding contextually appropriate responses to conflict for the groups with which you work.

Bottom Right: What Is Important to Us About Fast and Fair Conflict Resolution?

When you are working with a group to explore its values and overarching goals in relation to conflict, it can be helpful to make a distinction between task conflict, which is often helpful for groups, and relationship conflict, which is almost always unhelpful in groups.[1]

Task conflict is when group members have different opinions about how best to execute a task. This might arise because members perceive value issues differently, or because they bring different types of expertise to a task, leading them to see different solutions. In some situations, task conflict can be very healthy in groups. Healthy disagreement can contribute to creativity, learning, and better solutions overall. Groups that experience no conflict, not even task conflict, are usually composed of highly avoidant members who are deferential to the leader. Healthy task conflict can be seen as a sign of a thriving, autonomy-supportive environment where people are encouraged to think for themselves and voice their opinions.

Relationship conflict tends to be much more emotional, and it involves tension, annoyance, and animosity between people. Negative and rigid evaluations and judgments of others fuel relationship conflict. Whereas task conflict tends to produce more effective group decisions, relationship conflict is usually associated with poorer group decisions.[2] Furthermore, relationship conflict usually creates disconnect between people as they avoid one another, leading to failures to transmit information and coordinate activity. Paul once worked with a surgical team, the members of which were locked in relational conflict to the extent that they weren't passing on information about the treatment and current condition of patients at the handover between shifts, creating extreme risk for patients. Relationship conflict also consumes a huge amount of emotional energy, distracting people from the task and causing them to waste time gossiping and disagreeing.[3]

But, of course, the reality is that task and relationship conflict are often intertwined. What starts as a disagreement about how something should be done can quickly become personal if people hear "I don't agree with you" as "I think you are a fool." This is particularly likely to happen when trust is low and, consequently, people are more likely to misconstrue disagreement as criticism.[4] To the extent that you can use all the other core design principles we've already discussed to build a safe and trusting environment within your group, you can reduce the likelihood of people hearing disagreement as criticism, as well as increase the willingness of people to speak up when they disagree because they know that disagreement will be less likely to blow up into emotionally charged relational conflict.[5] When group members feel more psychologically safe, they are more likely to address conflict in a collaborative manner rather than avoid it.[6]

Clarifying the distinction between healthy and unhealthy forms of conflict can help a group articulate what's important to them about how conflict is handled. Healthy conflict tends to be focused on tasks and includes legitimate differences of opinion, values, perspectives, or expectations. Unhealthy conflict tends to be more focused on labeling or judging individuals. It often involves people competing for scarce resources or power, conflicts between individual and collective goals, and poor communication patterns.

In the case of the pharmacy, introducing the distinction between task and relational conflict led one of the pharmacists to use the phrase "play the ball, not the man," which, while somewhat sexist, is a common saying in Australia, pointing to the need to focus on the task rather than playing dirty by attacking the person in a football match. We recorded this saying, along with a host of other values and overarching goals that the pharmacy members felt were important to them about conflict (see the bottom-right quadrant of figure 12.1). While the pharmacy employees wanted to make a safe space in which people could voice disagreements and be honest, they also wanted the pharmacy to be a fun place to work where customers were treated well and staff felt respected and that their perspectives were understood.

Core Design Principle 6: Fast and Fair Conflict Resolution 179

Figure 12.1. The collective matrix applied to exploring processes for conflict resolution.

Bottom Left: What Difficult Thoughts, Emotions, Sensations, and Images Show Up and Get in the Way of Effective Conflict Resolution?

Almost everything that tends to show up in the bottom-left quadrant when we discuss conflict is related to fear. People are often afraid of conflict. They may have experienced highly destructive conflict in family relationships at a very young age, which left them feeling threatened and unsafe. Or they may have made honest attempts to voice their perspective only to have it misinterpreted and blown up into something much larger than they were expecting.

In our experience, talking about conflict often shows up in a bodily way for people, with some people literally becoming tense with fear or even locking up, crying, or getting angry at the facilitator for even raising the issue. If this seems likely to happen, it's useful in advance to teach the group some of the psychological flexibility skills we discussed in chapter 5 so that they can better self-regulate when discussing conflict.

Aside from fear of conflict, people often express judgments about others in this quadrant. Perhaps they see others as untrustworthy or too incompetent to deal with conflict effectively. Or they might just think that others are purely self-interested and don't have the good of the group in mind.

Finally, we often find self-focused thoughts in this quadrant. Will I have the skill to handle the conflict? Will I look foolish to others? And so on. In the case of the pharmacy, all of these issues were raised in the bottom-left quadrant of their collective matrix (see figure 12.1).

Top Left: What Do We Do to Avoid or Control the Difficult Experiences About Conflict?

The overwhelming strategy most people use to manage the difficult feelings and thoughts associated with conflict is to simply avoid having it. Husbands and wives come to an uneasy truce, knowing where the landmines are placed and skilfully avoiding talking about them for decades. Work teams learn what is "undiscussable," whether it be the manager's bad behavior, the person free-riding on the efforts of the rest of the team, or the lack of feasibility of a poorly thought through strategic plan.

Some groups institutionalize avoidance, by creating rules that all conflicts must be handled by the HR department, for example, terrified as they are that their organization will be sued by a disgruntled employee. Other groups insist upon external mediation or legal services to resolve conflicts that could easily be handled much more effectively and efficiently by group members, if they had some basic conflict-resolution skills.

If conflict is not dealt with effectively, it can undermine all the other principles. It is hard to maintain shared identity and purpose (core design principle 1), equity (core design principle 2), inclusive decision making (core design principle 3), transparency (core design principle 4), or effective responding (core design principle 5) when people don't trust one another enough to grapple with their differences.

Conflict can even undermine a group's capacity to manage itself using the first six principles (core design principle 7). One fascinating study found that self-managed teams tended to literally restructure themselves in response to unmanaged conflict. Over time, people shared less information with one another, were less trusting of giving others autonomy to make decisions, and avoided others with whom they were supposed to be cooperating, all of which eventually undermined many of the benefits of giving a group authority to manage itself.[7] See the top-left quadrant of figure 12.1 for an example.

Top Right: What Do We Want to Do to Manage Conflict Effectively?

When generating solutions that might populate the top-right quadrant of the matrix, it can be helpful to consider the three levels (mentioned earlier) at which change can occur:

1. Build interpersonal skills such as listening well and having integrative win-win conversations.

2. Enhance personal skills such as emotion regulation and perspective taking.

3. Create group-level agreements and processes for managing conflict efficiently and effectively.

In the pharmacy case, the group used the top-right quadrant to prioritize one of each of these examples (see figure 12.1). In the following sections, we begin with the idea of fostering integrative approaches to conflict and then explore the personal skills and group-level agreements needed to support such an orientation.

Interpersonal Skills: An Integrative Approach to Conflict

Paul recently facilitated a session between a political candidate running for office and his campaign manager. The candidate was gregarious, outgoing, and ambitious. He had big and admirable plans for what he wanted to achieve in office. But from the campaign manager's point of view, he was a risk taker, somewhat careless with the rules and processes of the party, self-aggrandizing, and a bit of a loose cannon. On the flip side, the manager was well organized and passionate about seeing all his party's candidates succeed in the next election (if they could only avoid any electoral missteps that would land them in the media for all the wrong reasons). But to the candidate, the manager was overly cautious and bureaucratic. What's more, the candidate was beginning to feel that he was not being treated fairly, with the party giving more resources and attention to other candidates. While their differences might have started in task conflict over what to do during the campaign, they had morphed into judgments and evaluations of each other's personality. The manager saw the candidate as untrustworthy, and the candidate saw the manager as obstructive and unfair. This conflict was having real impact upon the campaign, as their disagreements had become increasingly public and strident.

Paul was invited to help create a more positive and cooperative relationship between the candidate and the campaign manager. He began by talking to both sides separately, using the matrix to create a map of the substantive issues and all the associated judgments and assumptions on both sides. He then organized a three-way meeting.

Paul sat down with the two of them and a sheet of paper. He first established some ground rules for turn taking and reflective listening, establishing a mandate for himself to direct the disputants to reflect back what they had heard at key points if emotions started to run high. Once that was done, he quickly drew the matrix, without labeling the axes, and dived in with the question "What matters to you most in this situation?" Taking turns jotting down the priorities of each person, it rapidly became clear that they shared a lot in common. They both wanted the candidate to win the election, and they both wanted their political party to be in the media for the right reasons. But, whereas the manager's perspective was focused on avoiding risk for the good of the whole party, the candidate's energies were focused on making a big impression with his own campaign.

What would you do next if you were facilitating this exchange?

Before we explain how Paul approached the situation, let's pause to consider the types of conversations needed to solve "me versus us" and "us versus them" conflicts. Solving such problems invariably involves creating integrative, or win-win, conversations that benefit self, other, and the collective as much as possible. An integrative approach is about neither winning over nor giving in, neither competing nor entirely accommodating. It is about bringing one's authentic interests to the table while also listening carefully and mindfully for the authentic interests of the other (be it individual or group):

1. Separate the people from the problem.

2. Focus on the shared interests of both parties.

3. Develop many options that can be used to solve the problem.

4. Evaluate the options using objective criteria.[8]

The matrix provides an ideal tool for implementing these four recommendations and creating more integrative solutions. Having mapped out the values and overarching goals espoused by each participant, and having recognized that they had much in common, Paul was able to create enough of a sense of shared purpose (get the candidate elected) to begin exploring the fears and concerns that each had about the other's behavior and entering those in the bottom left. While this started at a fairly superficial level (I am concerned I won't be able to support his campaign fully), the tenor of the conversation changed when Paul said to the manager, "It sounds as though you don't really trust Charles [the candidate]." After a long pause, the manager admitted that was true. The candidate was a bit taken aback, but by now there was enough trust in the process and their shared purpose that he didn't feel as though he needed to defend himself. In effect, having a map of the interests and concerns of both disputants helped both candidate and campaign manager separate their goals and values from their judgments about each other's personalities, thereby separating the people from the problem.

Whereas a position states a specific action or event that needs to occur (top right), an interest is a broader value (bottom right) or concern (bottom left). Positions correspond to just one path of action, whereas interests leave open the possibility of many different paths to satisfy those interests. When making specific recommendations for action conflict, it's possible to think more abstractly about shared values to see whether there's a way to come up with a specific action that will meet the needs of both sides. In this case, how can we *both* have a vibrant, flexible, and dynamic campaign while also adhering to the rules and minimizing the risk of negative exposure in the media? Having established a shared purpose, it was much easier for both disputants to recognize that both of these overarching interests were important and could be integrated by looking beyond each person's positions.

Paul then moved to exploring the top-left quadrant (what they do when hooked) and the top right (what they wanted to do instead to be more in line with their shared purpose) and, in so doing, developed many options that might be used to solve the problem. As he wrote down possible solutions in the top-right quadrant (for example, regular check-in meetings to evaluate progress and raise issues and concerns, guidelines for agreed-upon use of resources, serious public endorsement by the campaign manager of the candidate's campaign), both the candidate and the campaign manager were voicing their concerns, which Paul was able to add to the bottom left. With a list of options for possible actions they could take, they were able to evaluate each option against the relatively objective criteria of the shared values and purposes that they held.

Once the candidate and the campaign manager were able to step back from their judgments they were able to see how each benefited the other, creating a win-win by aligning personal and collective values. Once the candidate realized that the campaign manager was sympathetic to his cause, he was able to back off and make room for the idea that other candidates also deserved the party's support for the good of the party. And once the campaign manager was able to see the energy and passion that the candidate was bringing to his campaign, he was able to agree to fully support the candidate just as much as he was supporting the other candidates.

When your group members seem to be stuck in conflicting positions, use the matrix to help identify the underlying interests in the bottom right (values and overarching goals) and bottom left (concerns—usually self-protective). To get from position to interest (both in ourselves and others), you might ask questions like these:

- What is most important to you about getting X [whatever they have stated they want]?

- What matters most to you here?

- What would you like to accomplish by getting X?
- What is it about this that really matters to you in the longer term?
- If you got what you are asking for, what interests and needs would that satisfy?

When positions are in conflict, shifting to the level of interests often creates a sense of shared priorities, common ground from which dialogue can proceed. Focusing on interests also tends to shift the focus from the unchangeable past or present to what we want for the future, which is the only place where changes can be made.

Just knowing about the integrative approach to conflict isn't enough to do it well. It's also important to spend some time with your group developing the necessary skills members need to listen well to one another, convey one's perspective as one's perspective rather than the truth, and bring the conversation back to purpose when it gets bogged down in relational conflicts. While there is insufficient space here to provide a course in communication skills, there are some fantastic resources for learning how to have better conversations in groups, particularly in the presence of conflict. Furthermore, we have written a Prosocial guide to difficult conversations ("How Can We Make Conversations More Prosocial?"), available at our website (http://www.prosocial.world), that you can use as an addendum to this book. You can deepen all this interpersonal work by including in your Prosocial process some work on the personal capabilities of psychological flexibility and perspective taking, to which we now turn.

Personal Skills: The Power of Psychological Flexibility and Perspective Taking

As we mentioned previously, many people have learned to fear and avoid conflict. And this fear is transmitted through the generations, as children learn how to respond to conflict in their most formative years. All families involve conflict, and if we've been exposed to situations as children where conflict meant real threat of social exclusion, punishment, or even physical violence, then our reactive patterns toward conflict are likely to be deep, powerful, and automatic.

Our patterns of avoiding and overreacting to conflict are toxic for groups. Like an old married couple no longer able to talk about its disagreements, but still harboring resentments, a team that does not engage with conflict is likely to be wasting effort on suppressing and avoiding what needs to be confronted. When people fear conflict, what they really fear are strong negative emotions, such as anger, sadness, and shame.

PSYCHOLOGICAL FLEXIBILITY

The psychological flexibility skills you already learned about have been shown to be particularly important for creating an environment in which conflict is successfully managed in groups. For example, a range of studies have shown that learning the mindfulness skills underpinning psychological flexibility can lower the tendency for people to behave aggressively and increase the likelihood of more prosocial behaviors.[9] One reason why this occurs is because people are less likely to react defensively to threats when they're more able to notice and accept their emotions and habitual patterns of responding.[10] Furthermore, learning to take the perspective of others, empathize with them, and tolerate the pain of conflict helps transform the meaning of conflict from something to be avoided into something to be used for learning and growth.

It's much easier to be psychologically flexible in integrative, win-win group cultures. The mentality of scarcity and competition generated by win-lose group cultures is both born out of, and contributes to, fear and anxiety. Fear feeds win-lose, and win-lose feeds fear. The more afraid I am of loss, the more I attend to the possibilities of loss. And the more I do that, the more afraid I become. It's a vicious cycle that drives group members apart. Building the psychological flexibility skills of the group, the capacity to act in the direction of what really matters even when it feels uncomfortable, is one component of building the personal skills of team members to manage conflict. Another component is perspective taking.

PERSPECTIVE TAKING ON SELF AND OTHERS

Abraham Lincoln once engaged in a fascinating piece of perspective taking. A woman who objected to his conciliatory speech regarding the Civil War accosted him and said, "We should be destroying our enemies." He said, "Madam, do I not destroy our enemies by making them our friends?" This is the power of fully embracing the other's perspective. But, of course, to do so we need to get ourselves out of the center of the picture. How are we to do that?[11]

Heifetz[12] uses the metaphor of "getting up on the balcony" for perspective taking. From the balcony one can see more of the dance floor than is possible when one is on the floor dancing; one can see relationships not visible from the dance floor. Getting up on the balcony is the "observing self" stance of mindfulness training.

One aspect of getting up on the balcony is to separate the people from the problem. In every conversation there are two tracks—content and relationship—running simultaneously, each with its own level of interests. We have interests in expressing ourselves, being right, and convincing others, but we also usually have interests in preserving and enhancing relationships, giving and receiving positive regard, and trusting and being trusted. When conversations get difficult, particularly when there is disagreement or strong emotion, it's easy to see the problem as lying with the other person, with their character or their identity. This escalates conflict as people see themselves being attacked or threatened.

A more helpful stance is to objectify the problem rather than the person, and to assume that everyone in the relationship has understandable intentions, at least from their own point of view. From this perspective, we can stand side by side instead of being squared off against one another, and thus work on solving the problem together.

Perspective taking doesn't just involve the cognitive demands of observing oneself operating alongside others who are operating from their own perspective; there are also emotional demands. How can we react with respect and inclusiveness toward positions that we find inaccurate, ill-informed, or even morally repugnant? If we cannot manage this, we simply erect barriers, which will make it harder to collaborate.

Unfortunately, the very people who most need to take the perspective of others—leaders—are less likely to do so when there are strong expectations that the job of the leader is to control. There now is good evidence that increased power leads to increased distancing from others, as well as a reduction in the capacity to take the other's perspective. Guinote and colleagues conducted a comprehensive review of how power affects people, including how it decreases their perspective taking and makes them treat others more as objects.[13] Devolving power to more people has the potential to have a positive impact on relationality and perspective taking.

Getting up on the balcony and taking a bigger perspective on our situation is a structured, learnable skill. We can be taught to get better at perspective taking. Let's explore some practices for doing this.

Tools for flexibly shifting perspectives: Sometimes we use exercises to encourage people to shift between seeing the world from other people's points of view. For example, we might invite them to imagine a situation where they were extremely annoyed with someone. We have them visualize the scene, including as complete a picture as they can manage of what the other person said or did. We also invite them to notice what they felt like doing in response to the other person.

Then we invite them to write down five different reasons why the other person might have behaved the way they did. We encourage them to write only reasons that the other person might potentially recognize and endorse—that is, to fully inhabit how the experience might make sense from the other's point of view. For example, "because they are unkind" is unlikely to be something the other would endorse, whereas "because they were afraid that..." may well be something the other would be willing to endorse.

It's very easy to get stuck in the notion of "being right" even when it's damaging the relationship, or at least leading to a kind of blindness to the point of view of the other. To counteract that, it's helpful to recall that we need to develop the capability to very actively put ourselves in the shoes of others. This can involve a variety of specific behaviors, such as these:

- Imagining what others might be noticing, feeling, thinking, and wanting
- Actively asking others what they're noticing, feeling, thinking, and wanting
- Listening reflectively to test understandings
- Reversing roles with others, perhaps exploring what it's like to argue for their position and interests

Using the matrix to shift perspectives: In situations where conflict is an issue, you can have people imagine someone with whom they are in conflict, and then complete an individual matrix from the perspective of the other. Typically this exercise not only reveals new perspectives about why the other person might be acting the way they are, it often increases a sense of connection to the other, thereby increasing the possibility of finding shared purpose in the midst of conflict.

Empty chair: Another useful technique is to use a physical prop, such as an empty chair, to symbolize an alternate perspective. If one person is unwilling to talk, it's sometimes possible to imagine what they might say or do by signifying their presence with an empty chair, thereby allowing some unilateral progress to be made with perspective taking. Groups can also use a chair to signify the collective group purpose. This necessitates the creation of a clear group purpose, and the chair serves as a continual reminder for individuals to take the collective into account in the decision-making process. Groups can also use the empty chair to symbolize the perspectives of stakeholders from the past, present, or future, or even different sides of a tension between ideas or goals. Such conscious use of the incredible human ability to shift perspective in person, time, and space is at the very heart of complex cooperative behavior.

Agreed-Upon Principles and Processes for Conflict Resolution

Aside from developing personal and interpersonal skills to manage conflict effectively, having agreed-upon principles and processes that everyone can trust can greatly help the process of conflict resolution.

There are two main pieces to an effective set of guidelines for conflict management: principles and processes. At a minimum, groups must do conflict management in a timely and fair manner (as core design principle 6 suggests). However, other principles, such as seeking to resolve the conflict "close to the source" first without involving external parties, focusing on the issue and not the personalities, distributing equally the power for anyone to escalate the process, or seeking to use listening and dialogue skills whenever possible, might serve as useful guidelines for shared practices. We offer a tremendous example of conflict-escalation agreement that incorporates all of these elements at our website (http://www.prosocial.world).

Conflict-management processes typically involve the idea of escalation according to need. Typically, effective conflict resolution begins as an informal conversation between the parties involved. If participants have learned some of the skills of emotional self-regulation, perspective taking, listening effectively, speaking assertively not aggressively, and holding a conversation to a clear purpose, it's often possible for them to simply work through differences in open conversation. Beyond that initial conversation, it's quite common to encounter conflict-escalation processes that move through stages as required, from one-on-one conversation to a mediated conversation to some kind of informal arbitration process to an external arbitration process, for example.

Conclusion

Conflict is the most obvious expression of tensions between interests within and between groups. Groups and groups of groups can have interests that are different from their constituent parts, including those of the individual. When members get good at the interpersonal and personal skills discussed in this chapter, and groups create group-level agreements about what to do when conflict emerges, a sense of safety is established, allowing members to express themselves honestly.

In this chapter we used the collective matrix to organize our thinking and recommendations regarding how to create more effective, timely, and fair conflict resolution in groups. We demonstrated how task conflict, not relational conflict, is not only ubiquitous in groups but can also be helpful. If groups successfully manage task conflict so that it doesn't become relational, they can improve not only the level of cooperation within the group but also the levels of creativity and performance. Another major theme we hope you take from this chapter is that conflict is often associated with fear and avoidance, and groups can improve their ability to resolve conflict if members develop psychological flexibility, which allows them to be open, aware, and connected to their purpose and the workability of their behaviors.

Finally, we explored the interpersonal, personal, and systemic aspects of creating a win-win, or integrative, approach to conflict resolution. An integrative, collaborative orientation does not require sacrificing one's own needs; they are just as important as anyone else's and need to be taken into account. Furthermore, being able to decenter from oneself as the center of experience, and focusing on the quality of any relationship and its workability for achieving purpose, increases the possibilities for different forms of relating and the likelihood that others involved in the conversation will engage and participate.

In the next chapter, we shift from looking at relations between members of a group and focus instead on a quality of the group as a whole: its authority to govern itself.

CHAPTER 13

Core Design Principle 7
Authority to Self-Govern

In 1996, a child riding a bike was struck by a car in a suburban neighborhood of Portland, Oregon. The accident was not fatal, but it brought the neighbors together to discuss what could be done to make their streets safer. One of the neighbors had lived in Italy and had fond memories of the wonderful piazzas there that provide safe public spaces for people of all ages to socialize and play. Another neighbor related the strong sense of community that he had experienced living in a Mayan village in Mexico. Inspired by these examples, the neighbors decided to transform the intersection where the accident occurred into a public space, which became known as Share-It Square. They began by painting the intersection, turning it into a colorful horizontal mural. A community bulletin board, minilibrary, food stand, kid's playhouse, and tea station were soon to follow.

But there was a problem—all of this was illegal! The upwelling of community spirit ran afoul of all sorts of rules and regulations the city of Portland imposed on its neighborhoods. The city was not trying to be repressive, but the entire culture of municipal governance in America is not designed to foster the kind of initiative that the residents of Share-It Square wanted to take. After trying and failing to work within the system, the residents ultimately decided to proceed, even if it required breaking the law. In fact, their rebellion added to their sense of identity and purpose. Over time, they prevailed and new ordinances were passed to foster rather than inhibit such community-building efforts. An organization called City Repair was formed that has now led to similar projects all over the world.[1]

This story illustrates a point that has tremendous repercussions for the successful implementation of the Prosocial process. For a group to become structured according to the core design principles, it must have a degree of authority to manage its own affairs. As obvious as this point might seem, groups violate it again and again in modern life. How many schools do you know where teachers are enmeshed in so many policies and procedures that they cannot actually manage their classrooms? How many government agencies are tied up in red tape? How many military units follow a strict chain of command? Even science-based solutions suffer from this problem, such as when a randomized controlled trial leads to a "best practice" that is replicated in a cookie-cutter fashion without regard to local context.

When Elinor Ostrom studied common-pool resource groups, the need for local authority emerged as so important that she made it a core design principle in its own right. By her reckoning, it was essential that external authorities not challenge the right of a group to form its own rules and impose its own sanctions, as long as they don't violate basic human rights. She encountered many examples of groups that were capable of managing their common-pool resources but failed for lack of sufficient local autonomy.[2]

Core design principle 7, authority to self-govern, is about the group, not the individuals within the group. Specifically, it's about the group's capacity to manage its own affairs without excessive interference from outside. "Authority" can mean the power or right to give orders, make decisions, and enforce obedience. But it can also mean the capacity and power to author one's own experience. We mean it in this latter way.

In a sense, this principle is the same as core design principle 3 (fair and inclusive decision making), but at the group rather than individual level. Both core design principle 3 and core design principle 7 involve rights and responsibilities associated with autonomy—the right to have a voice and the responsibility to put forward one's perspective and expertise. To the extent that either individuals (core design principle 3) or groups (core design principle 7) are incapable or unwilling to contribute, their interests are less likely to be satisfied, leading to decreased cooperation and effectiveness. And the barriers to executing core design principle 7 are in some ways like the barriers to executing core design principle 3. Senior leaders often fear that if they empower individuals within groups, or groups within groups, they will lose control and create chaotic decision making that is not aligned at an organizational level. In this chapter, we will explore three examples of how groups have managed these legitimate concerns to coordinate activity between as well as within groups.

An appreciation of local autonomy is not entirely lacking in modern life. In fact, it is reflected to a degree in social institutions that have culturally evolved over centuries and millennia. The concept of subsidiarity arose within the Catholic Church as "the principle that a central authority should have a subsidiary function, performing only those tasks which cannot be performed at a more local level."[3] The political concept of federalism follows the same principle, according a degree of authority to smaller units nested within larger units. Nevertheless, many institutions that claim local autonomy for themselves deny it to the groups that they preside over. A corporation might fight against government regulations but itself have a top-down command-and-control organization. A city or county might have its own authority in some domains, but become an obstacle for groups that form within its boundaries, as the residents involved with Share-It Square discovered, to their frustration.

That's the bad news. The good news is that energy and creativity can be unleashed when larger-scale entities, such as governments, corporations, and even the military, make it easier rather than harder for groups to form and manage their own affairs, and in a way that also contributes to the welfare of the larger-scale entity. In this chapter we'll tell three stories about larger-scale entities that successfully accomplished this transformation. But first we need to make a crucial distinction between functional design principles and their implementation.

Functional Design Principles and Their Implementation

When people first learn about the core design principles, they often think they sound like a cookie-cutter approach to managing groups. *Eight simple steps toward a better group!* But implementing any given design principle is not necessarily simple, and what works for one group might not work for another. A design principle such as fair and inclusive decision making or monitoring agreed-upon behavior tells you *what* needs to be done in functional terms, but not *how to do it*.

David learned this for himself when he was asked to help design a school for at-risk students in his hometown of Binghamton, New York.[4] This was just after he had met and started working with Elinor Ostrom, so he was excited to use the core design principles as a blueprint for the design of the school, along with two auxiliary design principles important in the context of learning. The first of these was to create a safe and secure social environment, since fear is not a state of mind conducive to long-term learning. The second was drawn from the kind of learning principles discussed in chapter 4: to make long-term learning outcomes rewarding over the short term, since nobody learns well when all of the costs are in the present and all of the benefits are in the future. Eager to use state-of-the-art assessment techniques, his team recruited twice as many students as the school could accommodate, randomly selected half of them to admit, and tracked the other half as a comparison group as they experienced the normal high school routine.

Everything looked great on paper, but when school actually started, it became clear that some of the plans for implementing the core and auxiliary design principles simply weren't going to work. They had to improvise on the spot. This experience is not unusual, and it doesn't reflect poor methodology. The core design principles are so sensitive to context that nearly every group will need to continuously monitor how its implementation is working, and change it as problems are encountered.

This is why it is so important for groups to have the authority to manage their own affairs. No one else is in a position to decide upon the best implementation. Beware of *any* behavior-change technique that insists upon a particular implementation! Only functional principles can be general.

Once you appreciate the need to continuously monitor and adjust your implementation of the core design principles, you'll see that we intend for you to review the principles continuously. The same is true of the matrix: it's something to be used and reviewed continuously. Here's how Robert Styles, an expert Prosocial facilitator, describes his approach to using the principles and monitoring their effectiveness:

> My sessions deal with the real work of the institution, the living responses of people in teams to the demands of daily life at work. In these sessions we explore which Prosocial principles underpin the particular phase of the program we are engaged in and debrief their experience of trying new things with their teams.[5]

With this in mind, here are examples of three larger-scale organizations that achieved amazing results by granting authority to self-govern to the groups they preside over.

Health Care

Everyone knows that health care in the United States has become a quagmire. Even in more enlightened nations, health care has become a dehumanizing bureaucracy. Against this bleak background, a Dutch health care system called Buurtzorg provides a shining counterexample that is spreading rapidly within the Netherlands and beyond.[6]

Buurtzorg is Dutch for "neighborhood care," and indeed the organization considers neighborhoods as the units of care. Each neighborhood is tended by a team of nurses, who take the time to get to know the residents as individuals and families. Buurtzorg recognizes the following human values as universal:

- People want control over their own lives for as long as possible.

- People strive to maintain or improve their own quality of life.

- People seek social interaction.

- People seek "warm" relationships with others.

The mission (core design principle 1) of the team of nurses is to help residents manage their own lives, and the team itself is given an extraordinary amount of freedom to do its job (core design principle 7). The teams are composed of twelve members who are responsible for taking care of their patients, as well as for managing the team's work and deciding how they'd like to be organized, make decisions, and share responsibilities. The office of each team is in the neighborhood it serves, and members are responsible for introducing themselves to the local community, general practitioners, therapists, and other professionals. Teams build caseloads through referrals and by word of mouth.

Especially for Americans accustomed to a depersonalized health care system, the idea that a nurse might come to your home and establish a long-term relationship with you, tailored to your needs and desires, might seem too good to be true. Wouldn't such a health care system be hopelessly expensive to maintain? No, because more time spent with residents in this way saves time and money elsewhere in the system. When residents become more autonomous, they require less care, including fewer emergency hospital admissions. When they do need to be admitted to the hospital, their average stay is shorter. According to the professional services firm Ernst and Young, the Dutch health care system would save 40 percent if all care were provided the Buurtzorg way.[7]

Giving each team so much autonomy might seem like a recipe for chaos, but there is actually a lot of structure to the teams and the network as a whole. The structure has evolved—and continues to evolve—on the basis of the teams experimenting, retaining their best practices, and transmitting their best practices to other teams for their consideration. As a result, Buurtzorg teams tend to score high on all of the core design principles—without having learned them from Elinor Ostrom, to our knowledge—simply because they discovered for themselves what worked.

As an example, here is how team members make decisions (core design principle 3). First a member is chosen to play a facilitator role. Then the agenda for the meeting is put together on the spot, based on what is most pressing to discuss. The facilitator is not allowed to make any statements, suggestions, or decisions and is only allowed to ask questions, such as "What is your proposal?" or "What is your rationale for your proposal?" Proposals are listed on a flip chart to be reviewed and refined. A full consensus is not required for the final decision. It is enough that nobody has a principled objection, with the understanding that the decision can be revisited in the future on the basis of experience and other additional information. One can easily imagine the many, many decisions made by teams that led to this pragmatic arrangement!

Buurtzorg is worth learning about and emulating in many respects, but for the purposes of this chapter it provides an outstanding example of the positive cultural evolution that can take place when groups are provided with local autonomy, especially within a larger and appropriately structured social organization.

Business and International Aid

In the 1980s, a business consultant named Robert H. Schaffer was present when one of his clients, a New Jersey oil refinery, experienced a wildcat strike. About 450 supervisors, managers, and engineers were forced to run an operation that normally employed a workforce of three thousand—and not just for a few days, but for four months. Not only did they manage to pull off the miracle, but they experienced intense pleasure doing it, despite the emergency situation being dysphoric in other respects.

Intrigued, Schaffer wondered whether this kind of peak performance could be elicited under normal working conditions. He experimented with creating "emergency" situations by challenging small teams within a company to accomplish daunting goals in a short amount of time, such as doubling the number of customers for a new product line in one hundred days. He discovered that the teams indeed leapt into action and displayed the same kind of zest one might see in an emergency situation. Hence, the concept of a "rapid results cycle" was born.[8]

It turns out that when lower-ranking employees most closely associated with a given challenge (such as increasing sales or customer satisfaction) are tasked with solving the problem, they come up with better solutions than those of top managers or outside consultants. Why? Because they are in a better position to understand the nature of the problem and devise workable solutions. The "emergency" atmosphere also allows bureaucratic rules to be relaxed. And group members are able to bask in their success rather than have all the credit go to their bosses.

Core design principle 7 is key to the success of rapid results. Large companies are highly complex systems with many parts that must work in a coordinated fashion to function as a whole. The whole system cannot be optimized by trying to separately optimize each part, and nobody is smart enough to anticipate all of the interactions among the parts. Accomplishing positive change in a large company is therefore a formidable task, and top-down efforts are more likely to fail than succeed. In part, because of their bottom-up nature, changes produced by rapid results cycles are small enough to be integrated with the larger business operation. Far from nibbling at the edges of fundamental change, rapid results cycles can become an engine of fundamental change in an incremental fashion, producing short-term benefits along the way without requiring expensive outside consultants. This combination of benefits might seem too good to be true, but it has been documented repeatedly, and some major corporations have adopted rapid results cycles as their main change engine.

It is important to stress that core design principle 7 alone does not account for the success of rapid results. Nothing creates a strong sense of identity and purpose (core design principle 1) like an emergency situation, whether real or manufactured! And the rapid results cycle requires *both* a bottom-up process (the rapid results teams) *and* a top-down process (a strategy for employing the teams in a way that results in a long-term systemic goal). Either process by itself would be inadequate.

After proving itself in the business world, the rapid results method spread to the seemingly very different world of international aid through the creation of a nonprofit organization called the Rapid Results Institute. This organization works with local populations to improve change processes. For example, previous efforts to persuade women to visit family planning clinics in Madagascar had resulted in gains of only a few percentage points over a fifteen-year period. The institute created rapid results teams composed of local women committed to the seemingly impossible goal of increasing the percentage by 30 percent in one hundred days. Not only did the groups meet their goals, but some achieved gains as high as 500 percent. One can well imagine the women springing into action in a way they would not have done before. More than a dozen developing countries are currently employing the rapid results method for problems as diverse as child malnutrition, HIV/AIDS, and corruption.

The Military

General Stanley McChrystal, former head of the Joint Special Operations Command, had a problem. The mightiest military power on Earth had occupied Iraq in 2003 without difficulty as a conventional military operation. Once there, however, they faced an enemy—Al Qaeda—that played by a different set of rules. As McChrystal put it in his book *Team of Teams*:

> The Task Force hadn't chosen to change; we were driven by necessity. Although lavishly resourced and exquisitely trained, we found ourselves losing to an enemy that, by traditional calculus, we should have dominated. Over time we came to realize that more than our foe, we were actually struggling to cope with an environment that was fundamentally different than anything we'd planned or trained for. The speed and interdependence of events had produced new dynamics that threatened to overwhelm the time-honored processes and culture we'd built.[9]

Coping with terrorist events that could take place at any place and any time required small elite teams that were authorized to make their own decisions on the spot without needing permission from anyone higher up in the chain of command. Their mission was to find, fix, finish, exploit, and analyze. *Find* refers to finding the target. *Fix* refers to finding its position in real time. *Finish* refers to removing the threat of the target on the battlefield. *Exploit* refers to searching the target (for example, a person or location) for any intelligence. *Analyze* refers to assessing the operation and making use of the intelligence gathered. Analysis could involve higher-ups and identify errors committed during a previous operation so they wouldn't occur in subsequent operations, but every other phase of the cycle could be undertaken by groups on their own.

The most famous teams that were part of the task force were the Navy SEALs. Most people imagine SEALs to be superhuman individual warriors, but McChrystal goes out of his way to stress that their training is much more about forming teams than it is about individual training. As he puts it, "The formation of SEAL teams is less about preparing people to follow precise orders than it is about developing trust and the ability to adapt within a small group."[10] Recruits who are in it for themselves are the first to quit. All of the training exercises are done in groups, and the physical privations are more about the bonding experience they invite than they are about physical training and endurance. By the time a Navy SEAL team is trained, its members combined truly compose a superorganism capable of acting in a coordinated fashion in a split second. It is thanks to McChrystal's team approach that the US military was able to fight Al Qaeda in Iran and Afghanistan as well as it did.

Using the Collective Matrix to Explore Core Design Principle 7

Core design principle 7 is all about giving the group the authority it needs to manage itself appropriately. The examples above illustrate how great things can happen when a group has freedom, or when its senior leader supports the idea of an autonomous environment. But what if you're in an environment where authority to self-govern is a problem? Essentially your options are limited to attempting to increase the group's authority to self-govern (which can sometimes be difficult or impossible), or attempting to work around the interference to reduce its impact on the functioning of the group.

If you look at this principle from a different perspective, it's about working against excessive outside interference. In our experience of working with this principle, problems with self-governance often stem from outside influences undermining or hampering core design principles 1 through 6. We've regularly seen this principle violated in government agencies or community groups that are reliant on government funding. In these groups the processes and procedures of top-down bureaucracy have grown to such an extent that it's almost impossible for work teams to manage themselves appropriately. Here are some examples we've seen:

- A government agency in which constantly changing government priorities and departmental structures left work groups unsure of what exactly they're responsible for, or what has the highest priority, both of which severely hampered the work teams' capacity to establish a clear sense of shared purpose and identity (core design principle 1).

- An agency in which it was impossible for workers to discourage free-riding (core design principle 2), independently check on or call out unhelpful behaviors (core design principle 4), respond to underperformance (core design principle 5), or resolve conflict (core design principle 6) because of restrictive HR policies that arose from fear of legal action and assumed that only HR professionals could deal with these matters.

- A publicly funded disability services group that was so hampered by the terms of its funding that it was unable to adapt its services to appropriately meet the needs of its client base (core design principle 1). This group's prevailing culture of risk avoidance quashed individual creativity and ideas to such an extent that staff members stopped making suggestions for improving services (core design principle 3).

In situations such as these, there might be room to lobby upward or outward for increasing authority to self-govern. We have seen groups make a case for why having more control over the execution of core design principles 1 through 6, as well as other internal processes, would enhance their capacity to achieve the aims of the group and of the broader system. This typically takes a strong group leader who is able to prosecute the case effectively.

We've also seen groups question whether the impairment of group functioning is actually as bad as it appears, and seek ways of working around the interference. Some groups and group leaders adopt an "act first and ask permission later" approach that assumes that actions taken in the best interests of the group will generally be forgiven even if they contravene policies imposed from outside. Of course, this can be a risky strategy, and it requires a clear sense of purpose, as well as group support, in order to work well.

It is possible to use the collective matrix, as illustrated in previous chapters, to work with a group to explore options for increasing authority to self-govern. Consider the case of the disability services group. Although we did not use the collective matrix with this group, we imagine that if we had the group might have clearly identified the reasons why increasing group autonomy might have been helpful, and perhaps it might have begun to uncover some lines of action that might have been helpful (see figure 13.1).

```
                                                    outside
3. What do we do to avoid or control                 ↑   4. What do we want to do to increase our
those difficult experiences in 2?                        authority to self-govern or manage the bits
                                                         we cannot self-govern?
Follow policy without really thinking about
what is in the best interest of clients                  Use the arguments revealed in the bottom
                                                         right to make a case for more freedoms
Stop making suggestions for change                       from the funder
Take on fewer clients
                                                         Recognize that outside constraints are
Give excuses to clients, blame the                       sometimes inevitable in such a large
government                                               system

                                                         Focus on what we can do rather than what
                                                         we cannot do

away  ←─────────────────────────────────────────────────────────────────→  toward

2. What difficult thoughts, emotions,                    1. What matters most to US about having
sensations, and images show up and get in                more authority to self-govern?
the way of effective conflict resolution?
                                                         Better services
"I feel helpless"
                                                         Increased opportunities to use our
"We just have to do what the funders                     knowledge
want"
                                                         Self-determination
"Doing anything else is risky"
                                                         Implementing good ideas relevant to us
FEAR of getting out of sync with other
groups                                                   Building group energy and efficacy

We might hurt someone                                    Fitting group activities to our strengths
                                                    ↓
We may not be delivering value              inside       Good decision making

It is hard work to manage ourselves, we
don't have the resources
```

Figure 13.1. A hypothetical collective matrix for a disability services group that identifies how the constraints of the government-funding agency are a significant block to group cooperation.

However hard it might be for a group to become more autonomous, the results are likely to speak for themselves—and the Prosocial process is exceptionally well designed to document group performance. If you're fighting for autonomy, try to get your "superiors" to let you try self-governance on a pilot basis. While you're at it, lend them a copy of this book! They might see your point more clearly when viewing it through an evolutionary lens.

Conclusion

We began this chapter by presenting a series of case studies documenting what happens when core design principle 7 is implemented well. It would be hard to imagine domains of modern life more different from each other than health care, business, international aid, and the military, but all of them benefited from implementing a strikingly similar small-team approach. By now the common denominator in all of them is probably very familiar to you. Success in each example called for a high degree of cooperation and adaptability, which small groups are exceptionally good at—but only if they are empowered with the authority to manage their own affairs! It is not an exaggeration to say that if this insight were more widely appreciated and adopted, the world would start operating more effectively.

Of course, all of these examples involve having enough power within the system as a whole to structure relations between groups. As the second part of the chapter demonstrated, if you are working within a context where your power is limited to your own group, increasing authority to self-govern can be a matter of "managing upward" to demonstrate the benefits of giving the group more autonomy. However, we don't deny the difficulty of such a strategy in many contexts, particularly those that are strongly hierarchical.

There is something else that our positive examples in this chapter share in common. Not only were small groups granted the authority to manage their own affairs, but they were also part of multigroup organizations that were organized in an intelligent way, what McChrystal calls teams of teams. This characteristic—collaborative relations among groups—is the final core design principle, which we describe in the next chapter.

CHAPTER 14

Core Design Principle 8
Collaborative Relations with Other Groups

Suppose we could wave a magic wand and cause all groups to function better by employing the principles we've outlined so far in this book. That would be a great leap forward for our world—but it wouldn't be enough to transform all groups, because modern life is a complex ecosystem with many human groups interacting with each other and the rest of the biological world. Unless we can orchestrate the interactions *between* groups in addition to *within* groups, we will not be able to solve the problems of our age.

The principles we've covered so far help solve the "me versus us" problem within groups, but if we're to scale this approach to enhance whole systems of cooperation, we need to address how "us versus them" can be reconciled. The key to the effectiveness of the Prosocial process is that it not only helps small groups work well internally, but it helps groups work well with other groups. The genius of Ostrom, and perhaps the biggest contribution of the Prosocial process, is that its principles are scalable to any level.

Multilevel selection theory casts the necessity of managing between-group interactions in particularly sharp relief. This theory makes it crystal clear that cooperation within groups cannot evolve without a process of competition among groups. Between-group competition need not be harmful. Darwin stressed that nature is not always red in tooth and claw. For example, a drought-resistant plant can outcompete a drought-susceptible plant in the desert without the plants interacting at all. Likewise, groups that function well can replace groups that function poorly just by surviving longer, giving rise to similar groups to a greater degree, either by direct replication or by imitation.

Nevertheless, just as nature is sometimes red in tooth and claw, cultural group selection sometimes does take the form of direct competition, as it has throughout human history. In these cases, group selection doesn't eliminate conflict so much as elevate it to the level of between-group interactions, where it can take place with even more harm than it causes at lower levels. There's another layer of selection involved in the evolution of cooperative interactions among groups, one that favors more cooperative groups of groups over less cooperative groups of groups. The general rule of multilevel selection is that adaptation at any given level requires a process of selection *at that level* and tends to be *undermined* by selection at lower levels.[1] Many social pathologies are examples of lower-scale cooperation that's disruptive for higher-scale social units. For example, cooperation among relatives can become nepotism. Cooperation among friends can become cronyism. Policies based on "My nation first" can be toxic for the planet.

Selection at the upper levels of a multitier hierarchy of groups might seem inherently more difficult than that at lower levels. Perhaps, but consider that each of us, as multicellular organisms, are already groups of groups of groups—in fact, we are multispecies ecosystems when you take our microbiomes into account![2] And our current societies are also groups of groups of groups that evolved through cultural evolution over the last ten thousand years.[3] We might complain that our nations and economies don't work well, but they would have been inconceivable to our distant ancestors. Thus, there are grounds for optimism that we can make the world more prosocial at the level of multigroup cultural ecosystems, in addition to single groups. That is what core design principle 8, collaborative relations with other groups, is all about.

Polycentric Governance and the Scale Independence of the Core Design Principles

In addition to her pioneering work on common-pool resource groups, Elinor Ostrom also thought deeply about governance at the multigroup level with her husband Vincent Ostrom and others. They called their school of thought polycentric governance,[4] and it was based on the following principles: (1) life consists of many spheres of activities, (2) each sphere has an optimal scale, and (3) good governance requires finding the optimal scale for each sphere of activity and coordinating among the spheres.

The concept of polycentric governance has the same quality of common sense, bordering on obviousness in retrospect, as the core design principles do for single groups. Yet, it is a far cry from the way that most multigroup societies are governed! As an example, Elinor Ostrom studied metropolitan police departments at a time when consolidation (bigger departments replacing smaller departments) was all the rage.[5] She showed that there was no single answer to the question of scale. A forensic lab could be regional, but it was important for cops on the beat to be more local so they could establish personal relationships with the people under their watch. One reason that police departments in the United States have found it difficult to act upon this insight is that there is far more federal funding available for militarizing police departments to fight the so-called wars on drugs and terrorism. Top-down financial incentives have crowded out the neighborhood policing policies that are actually needed—an example of large-scale governance that doesn't work out for the common good.[6] The bottom line is that polycentric governance does not spontaneously evolve—it must be mindfully constructed, similar to relations among individuals within single groups.

A central insight for polycentric governance done right is that the core design principles are *scale independent*. Polycentric governance offers a fractal view of groups as organisms in their own right, allowing conversations to shift to how those entities might implement principles 1 through 7 in their relationships with one another. Groups of groups also need their own shared sense of identity and purpose (core design principle 1);

they must distribute contributions and benefits fairly among their member groups (core design principle 2); all groups should take part in decision making (core design principle 3); agreed-upon behaviors should be monitored (core design principle 4); there should be praise for good behavior and graduated sanctions for deviant behavior (core design principle 5); there should be ways to resolve conflicts among groups that are fast and fair (core design principle 6); and the group of groups should have the authority to manage its own affairs (core design principle 7). The same can be said of groups of groups of groups—created appropriate relations at still larger scales, potentially to the scale of the entire earth.[7] In this sense, core design principle 8 is less a separate principle and more about the applicability of principles 1 through 7 at any level of organization.

All very well, you might be thinking, but how can core design principle 8 be implemented in a practical sense? We will outline three answers to this question. First, when a single group grows in size, it must differentiate its members as groups of groups in order to function well. Think about how many groups compose a school system, a corporation, a hospital, or a government agency. In these cases, there's a lot of scope for implementing polycentric governance for the whole organization.

Second, if your group or multigroup organization has become prosocial enough to function as a corporate unit, then you can work with like-minded groups to form prosocial consortia that thrive in competition with less-organized groups. A number of enlightened business corporations are taking impressive steps in this direction, as we will describe.

Third, groups often fail to implement the core design principles and polycentric governance because of competing narratives that seem to justify behaviors such as laissez-faire economic theory, which falsely gives the impression that the pursuit of lower-level self-interest robustly benefits the common good. As we discussed in chapter 2, the *Homo economicus* worldview destroys community by suggesting that focusing only on the interests of our own groups (families, organizations, or nations—for example, "Make America great again") is not only natural but preferable. This mode of operation is little different from that of a terrorist cell, which has intensive and successful cooperative relationships within the group but fails to cooperate with the purposes of the rest of society. The more this false and destructive narrative can be replaced by a more scientifically validated narrative based on multilevel selection theory, then the more the core design principles will "make sense" at all scales, including the scales of nations and the global village.

There are many forms of polycentric systems already in existence. For example, the polymath Michael Polanyi identified science, art, religion, and the law as all being polycentric in nature and driven by certain ideals, such as truth, beauty, transcendent truth, and justice.[8] Consider science. The world's scientists have no single centralized body that tells them what to do. Instead, they organize as individuals and teams globally, united by certain core values, such as replicability, and integrate through mechanisms such as journals, scientific organizations, and networks of universities. If you look for "the leader" of the scientists, you will not find one. But the system manages to be, on the whole, extraordinarily cooperative and well coordinated.

Multigroup Organizations with the Authority to Manage Their Own Affairs

Imagine a start-up company that begins with just a handful of people. The work is hard but very fulfilling because everyone believes in their product and is giving it their all. It's easy to implement the core design principles at such a small scale without even thinking about them. If the company is lucky enough to grow, then new staff members will need to be added. Roles within the organization will begin to differentiate. The single group will, by necessity, become a group of groups. Yet, because all of the subgroups exist within a single corporation, the corporation has as much authority to organize the interactions among groups as it does the interactions within groups. How well does the corporation do?

Just as single groups vary in how well they implement the core design principles when left to their own devices, from the best to the worst, the same can be said for multigroup organizations. At the worst end of the continuum, the sociologist Robert Jackall, in his book titled *Moral Mazes: The World of Corporate Managers*,[9] documents the unbridled competition that takes place among individuals, teams, and subdivisions within corporations, which creates chaos at the level of corporate performance. As a second example, consider Sears, once the largest retail store chain in the world, which declared bankruptcy due to its CEO's policy of pitting its divisions against each other in ruthless competition.[10]

At the best end of the continuum, all the groups of the three examples described in the previous chapter—Buurtzorg, rapid results, and the activities of the US Special Operations Forces under the direction of General Stanley McChrystal—function well at the level of the multigroup organization, in addition to the single-group level, largely because they created teams of teams. In the case of Buurtzorg, there is extensive communication between teams so that they can learn each other's best practices. Its founder, Jos de Blok, is so collaborative that he even serves on the board of competing health care services. His goal is clearly to establish quality health care worldwide, not his own wealth or even the predominance of his own company.

Because so many groups are divided into subgroups, Prosocial facilitators have their own experience working at the multigroup level. David worked with a National Public Radio station with twenty-eight employees who were divided into a number of departments called "verticals." Rather than hold meetings with the existing subgroups, David decided to meet separately with the managers and with small groups of staff members drawn from all of the verticals. It quickly emerged that the existing social organization was part of the problem. It was hard to have a strong sense of identity and purpose for the station as a whole (core design principle 1) because staff members didn't interact with people outside their verticals. Many felt that they had expertise to offer outside their verticals that wasn't being made use of. One step toward improvement, which was suggested as part of the process, was to post problems on a bulletin board located in the staff common area for all to see. More generally, the station decided to work toward a more fluid structure of teams organized around problems and opportunities—rapid results style—rather than rigid verticals.

Perhaps the Prosocial facilitator with the most experience working at the level of a multigroup organization is Jim Lemon, a medical pediatric psychologist in a district general hospital in Scotland. He describes his experience in an interview with Paul.[11] As with many Prosocial facilitators, Jim first became trained in acceptance and commitment therapy and used it successfully as a therapy and training method for individuals, but he realized that many problems could only be solved by working with various groups, such as the families of his patients and his own colleagues who weren't communicating with each other. As soon as he encountered the Prosocial process, with its focus on working with groups, he started to incorporate it into his own practice whenever possible. At first he worked at the level of single groups, starting with a pediatric cystic fibrosis team, which consisted of two consultant pediatricians, a dietitian, a physiotherapist, a specialist nurse, and Jim. The team was functioning moderately well but suffered from communication problems and burnout. It took a while for group members to open up to the matrix, which requires more self-disclosure than many people are accustomed to in a work setting, but soon completing it became second nature, and the matrix proved to be a powerful tool for them to create a strong sense of identity and purpose (core design principle 1) and to improve communication (core design principle 3). Something else that worked for Jim was to draw a contrast between what he called "the medical model" and evolutionary theory. The medical model is mechanical and involves doing tests, diagnoses, and treatments. Evolution is about a flexible process of variation and selection. Team members resonated with the evolutionary perspective, not as a replacement for the medical model but as a complement to it.

Once the Prosocial process proved itself useful with some groups, interest spread by word of mouth. This is how Jim put it in the interview:

> We knew from the start that if people had a good experience they would talk about it and other groups would want to start using the technique. This has happened, and it is spreading both through other medical teams and also to other professionals. For example, the physiotherapists found it so useful that they took it back to the other pediatric physiotherapists. Another example is that the dietitian also works in the gastroenterology section, so the ideas spread there. It just works out horizontally that way. I describe it as like a virus spreading wherever there is an opportunity. Medical teams love that metaphor.

Here is what Jim has to say about core design principle 8:

> The eighth one is useful to highlight because we work in so many different teams and with different levels of resources. We have lots of conversations about competing for resources; for example, if a pediatric service does well we are now in competition with a diabetes team for staff and services. Or we might be competing with the level above us, which might be the bigger units. There is a

history of tension between local smaller units and the bigger teams who are more [specialized] and ignore the local context. It is very easy for patients to fall through the cracks. We are seen as the locals down at the "Shire," and we see the big teams from "Mordor" as intimidating. There is plenty of scope for tension there, but talking about polycentric governance and collaborative relations between groups [core design principle 8] is very helpful.

Jim's description makes it clear that adaptive forms of polycentric governance are highly unlikely to spontaneously emerge in a health care system or any other multigroup organization. There are too many pressures pitting the various groups against each other, giving rise to lower-level identities, and so on. A mindful process of cultural evolution is required that keeps the well-being of the whole organization in mind, which is what the Prosocial process is in a position to provide to any multigroup organization with the authority to manage its own affairs.

Working with Other Prosocial Groups to Establish Prosociality at a Larger Scale

Suppose that our start-up company has successfully made the transition from a single group to a multigroup organization. Staff members are just as energized as the founders were, in part, because they operate in small teams that are given challenging tasks and enough authority to devise their own solutions. Like all companies, however, this one operates within a larger ecosystem that includes its suppliers, customers, and regulatory agencies. Also, staff members might have problems outside the workplace that impact their workplace performance. Ultimately, a sustainable human economy depends upon sustainable relations with the natural world. Just as it takes a village to raise a child, it takes a whole ecosystem to sustain every element within the ecosystem.

Knowing how well the core design principles work for the company, both within and between its groups, the company wants to cultivate similar relations with other members of the ecosystem—but its capacity to do so is more limited than it is for organizing its own groups. Still, *less* influence is not the same as *no* influence. The ecosystem is likely to contain at least some other organizations that are like-minded or can be educated. By cultivating relations with these members of the ecosystem, our company can help create a cooperative group-of-groups that can collectively outcompete less cooperative members of the ecosystem. This kind of between-group competition leads to an outcome that is benign rather than harmful—the implementation of the core design principles at a larger scale.

The New Belgium Brewing Company, whose motto is "business as a force for good," is a good example of a company that has done this. Here is how their 2018 "Business as a Force for Good" report starts out:

Business is one of the strongest forces in the world, and as New Belgium and many others have proved, it can be used as a force to do good. Our beer is made possible by well-functioning ecosystems and stable societies. We don't take that for granted, and we know we have a role in stewarding these. At New Belgium we have an entire team of coworkers dedicated solely to understanding and improving our beers' impact on ecosystems and communities. But the work is not limited to just one team. Our 700+ coworkers across the entire brewery make hefty contributions every day.[12]

Of course, many companies have public relations departments that pump out language such as this, just as some individuals falsely portray themselves as prosocial when they are not. In both cases, the fakes can be distinguished from the genuine article if there is enough transparency. And the fakes can fail in their deception if the genuine articles avoid or punish them. The New Belgium Brewing Company makes ample information available to verify that it walks the walk, rather than just talking the talk. Its own prosociality does not detract from its profits because it has formed connections with enough like-minded partners to thrive as the recipient of their prosociality.

Being a genuinely prosocial business doesn't just work by attracting like-minded customers. It also works by reducing costs, such as the making of beer, over the long term by creating an efficient ecosystem of suppliers and consumers of the waste products, such as those of the beer-making process. It might seem that any beer company would want to create such an ecosystem to increase the efficiency of its operation, but companies that are trying to maximize quarterly profits don't think or act that way.

New Belgium is one of thousands of companies that qualify as B Corps (where B stands for "benefit"), a designation similar to the Leeds certification for buildings. The certifying organization is called B Lab. Companies submit information about their policies regarding workers, community, customers, and environment, which B Lab uses to calculate the company's B score. If it is sufficiently high, they can advertise themselves as a B Corp and join the community of other B Corps to interact prosocially with each other—the formation of cooperative groups-of-groups in action! A recent study showed that B Corps perform equally well or better than a matched sample of standard corporations as far as profitability is concerned.[13] They truly do well by doing good.

B Corps comprise only one of many social responsibility movements in the business world. Another is the Business for Peace Foundation, which has established the equivalent of a Nobel Prize for enlightened business leaders and promotes the concept of being "businessworthy."[14] David attended and gave a seminar at the 2018 summit held in Oslo, Norway. In an interview he conducted, the foundation's founder, Per Saxegaard, said, "In a transparent, interdependent world, businesses wanting to sustain successfully will increasingly find it both natural and smart to behave ethically and responsibly."[15]

To summarize, prosocial groups can work together to create prosocial ecosystems in the same way that prosocial individuals can work together to create prosocial groups. That's what it means for the core design principles to be applicable at all scales. Why hasn't this been obvious to everyone all along?

The Problem of Competing Narratives

Nothing is obvious all by itself, but only against the background of other beliefs and assumptions. Almost everything we do has *some* rationale. Consider some educational policies in the United States, which include segregating students by age, extending the school day and year, eliminating recess and "frill" topics such as art to concentrate on the 3Rs (reading, writing, and arithmetic), and implementing no-touch rules to avoid potential sexual harassment. All of these are well-meaning policies that have a rationale, but they also have had terrible unforeseen consequences.

In general, problems occur when the beliefs and assumptions that give rise to our rationales fail to match up with reality, causing our best intentions to go horribly wrong, such as it did for the folktale character who was granted a wish, only to end up regretting what he wished for. Albert Camus, the novelist and existential philosopher, wisely wrote, "The evil that is in the world always comes of ignorance, and good intentions may do as much harm as malevolence, if they lack understanding… There can be no true goodness nor true love without the utmost clear-sightedness."[16]

One of the most destructive narratives of our time is that the exclusive pursuit of self-interest in a free-market economy robustly benefits the common good (see chapter 2). Here is how the economist Herbert Gintis describes its impact on the business world:

> After World War II, business schools blossomed all over the United States. All the major universities set up business schools. Before that, businessmen were just businessmen. They didn't go to college, or if they did they didn't learn anything about business. But these new business schools were very professional. When they wanted to teach economics, they simply borrowed from the economics discipline. In economics it's called *Homo economicus*. *Homo economicus* is not that popular any more but it certainly was after World War II. *Homo economicus* has no emotions. He's totally interested in maximizing his wealth and income. He really doesn't care about other people, although he does care about leisure. Leisure, income, and wealth are the only things. When they taught this to business school students it obviously followed that if you're a good businessman you should just maximize your material wealth. This is greedy. Being greedy is human, it's good to do, and the more greedy you are the more successful you'll be.[17]

This narrative reached its zenith when the economist Milton Freedman declared that the only social responsibility of a corporation is to maximize profits for its shareholders, and UK Prime Minister Margaret Thatcher declared that there is no such thing as society, only individuals and families. The very concept of society became nonobvious in her worldview! It's important to stress that the legions of people who are under the spell of this narrative are not evil or mean-spirited. For example, Alan Greenspan, who served as the head of the US Federal Reserve Board for decades and played a giant role in implementing neoliberal economic policy, was by all accounts a decent man trying to improve the common good. As Camus said, goodwill can cause as much damage as ill will if it's not enlightened.

It's because of this narrative of competition—buying into one depends on context and worldview—that we've made it such a priority in this book to convey the broad evolutionary worldview, rather than just describing the principles of the Prosocial process in a how-to fashion. One of the most effective ways to make the world more prosocial, at all scales, is to simply see the world through an evolutionary lens.

Using Prosocial at the Ecosystem Level

Much as we love working with single groups, we are even more excited about the prospect of working with many groups at a given locality to create prosocial ecosystems. You may recall Paul's work with the Canberra Innovation Network that we described in chapter 3. The challenge there was to use the matrix to map the individual interests of various organizations rather than those of individuals, and to try to bring those interests together in the form of a shared purpose and identity that supported Canberra's goal to become a center of excellence in innovation in Australia.

As we write these words, David is organizing an event in his hometown titled "Evolving the Future of Binghamton." The invitation begins, "Do you love the Binghamton area and want it to thrive? Are you frustrated by all of the obstacles that must be overcome, which might have thwarted your efforts in the past? If so, then join us for an evening of food, drink, conviviality, and *change*." The invitation is being sent to residents who are in a position to work with groups. Participants will be introduced to the Prosocial process and go through the group matrix with the whole Binghamton area envisioned as the group. Following the event, a team of Prosocial facilitators will work with a cohort of groups. Based on their experience, we expect the Prosocial process to spread virally, as Jim Lemon put it, and for the groups to work with each other to begin assembling a prosocial ecosystem, with additional facilitators being trained to keep up with demand.

Steve is taking the lead organizing Prosocial training in "boot camps" that can train more than one hundred facilitators at a time over a period of a few days. This approach is based on a model that already exists in the form of acceptance and commitment therapy (ACT) boot camps, intensive multiday training events held in major cities around the world. The wide availability of training is one reason why ACT is spreading so fast as a therapeutic and training method. Steve is currently exploring two models: adding a day of training in the use of core design principles with small groups to ACT boot camps that already include training in the matrix, as well as organizational work, or implementing boot camps focused exclusively on the Prosocial process, similar to what David is attempting in Binghamton.

If prosocial ecosystems begin to take root at different locations, then groups in these locations can start working together to implement cooperative efforts at larger scales. That is the beauty of a set of principles that can be applied at any scale, from the members of a small group to the nearly two hundred nations that make up the global village.

Conclusion

While Ostrom listed appropriate relations with other groups as an eighth principle, it is more usefully regarded as the application of the first seven design principles to another layer of a multitier hierarchy of groups. A group with a strong sense of purpose (core design principle 1), with mechanisms that allow for coordination and suppress disruptive self-serving behaviors (core design principle 6), and with local autonomy (core design principle 7) truly becomes a superorganism capable of extending prosociality to its interactions with other groups. This gives the core design principles approach tremendous generality. When one pauses to think about it, it's amazing that such a simple set of principles can apply to all kinds of groups and at all spatial and temporal scales.

Working in a multitier hierarchy might seem dauntingly complex compared to implementing Prosocial within a single group. Perhaps, but it is still possible, and there is no alternative! Many social units that we already think of as a single entity, such as a corporation, government, hospital, or school, are already a collection of groups with as much potential for managing intergroup relations as intragroup relations. And a prosocial group can work to improve its own social environment by banding together with other prosocial groups without asking anyone's permission. In our own efforts, while we are excited to work with single groups, we are even more excited to work with whole cultural ecosystems.

CHAPTER 15

Goal Setting for Action

The Prosocial process is about enacting change to make groups more collaborative and effective. Therefore, it is essential that setting concrete goals for action follow the discussions of the core design principles (CDPs). If you've used the collective matrix to explore the different CDPs, then you will have derived concrete goals from the top-right quadrants of the various matrices you've conducted. If you have simply discussed how to enhance the implementation of each of the CDPs, that process will have also given you a raw set of concrete goals. Either way, you'll need a process for compiling and prioritizing the goals of the group over the short and medium term. You will need an action plan.

Why Goals Matter: Performance and Learning Goals

Goals are important for groups because they help them coordinate action in a shared direction, and because they motivate people to initiate and sustain effort, even when the going gets tough. But, done well, goals can also serve another critical purpose: they can help people learn and adapt over time.

It's usually helpful to talk about the distinction between performance goals and learning goals with groups. Performance goals are focused on the outcome to be achieved: for example, "Create a shared purpose by May 16," or "Implement the new conflict transformation policy by June 23." Such goals are perfectly reasonable and helpful when everybody in the system knows exactly what to do and can predict how every other part of the system will react.

But we have repeatedly emphasized throughout this book how human systems are complex adaptive systems, evolving on multiple levels, with nonlinear cause and effect between agents and levels. As we become increasingly densely interconnected with one another, the extent of complexity also increases. Whereas twenty or thirty years ago solving problems of coordination and cooperation might have required simple problem solving and goal setting, now it more often than not requires a more learning-oriented approach, experimenting with trying out things to see how they work, and remaining open to flexibly changing direction in the face of feedback from the system.

A learning goal is focused less on what must be achieved and more on learning how to get to that outcome. In the context of the Prosocial process, learning goals often mean experimenting with different approaches to putting the principles into action, and continually revisiting both the effectiveness of those experiments and the unfolding attitudes, beliefs, and intentions of the actors in the system (using, for example, a tool such as the matrix). Some examples of learning goals include "Finding and comparing at least two different approaches to consent-based decision making that could be implemented in a school setting," or "Evaluating the effectiveness of our new conflict-resolution process for three months using a weekly check on critical incidents with all staff." You can see that although there is no formal difference between a learning goal and a performance goal from a behavioral science perspective (both involve planning to do something worthwhile), there is a difference in the feel of learning goals. Indeed, learning goals can help increase the psychological flexibility of groups. Learning goals help the group members hold the goals more lightly, recognizing the need to try things out (variation) and retain only that which is helpful (selection and retention). Learning goals help group members be more willing to make mistakes and less likely to seek to blame others if things go wrong. In other words, learning goals make the process of evolution in complex adaptive systems more visible.

Having done some work on psychological flexibility first can be a real asset when a group is setting learning goals. A group that remains mentally open to new ideas while resolutely focused on shared values and purposes is much more likely to learn its way through complexity than a group that is closed to new approaches or has lost sight of the reasons for its existence. Building psychological flexibility helps people deal with the discomfort of uncertainty and ambiguity, reducing the likelihood of relational conflict and increasing trust among group members. Think of a psychologically flexible group as a powerful and well-tuned car headed out on a trip. Driving toward a destination is where the rubber meets the road, but the functioning of the car is what most determines whether the group will reach its destination. A wrong turn or two will not prevent the group from reaching its destination, but if the engine breaks down and the wheels come off the car, the group may never get there.

Characteristics of Good Learning and Performance Goals

The literature on goal setting is large and well established with relatively consistent findings.[1] In this section we'll discuss the underpinnings of appropriate and attainable learning and performance goals. Learning goals are focused on acquiring and testing new strategies to attain valued outcomes, whereas performance goals are focused on the outcomes themselves. Both benefit from being as specific, measurable, attainable, relevant, and time-bound as possible.

Learning or performance goals that are specific are much more likely to lead to higher performance than vague goals, such as "do your best" or "try hard."[2] Even a moment of consideration makes clear why this is the case. People need specific goals because it's difficult to know when a vague goal has been reached. With vague goals the processes of variation and selection that are key to the evolution of effective systems are undermined by weak selection criteria. In examining a learning or performance goal that flows from the Prosocial process, the group facilitator should try to ensure that the group considers the many "w" issues involved, such as *what* needs to be accomplished, by *whom*, by *when*, and so on. Group members need to understand *why* the goal matters and *which* resources will be deployed *where* in that effort.

Measurable goals flow fairly directly from the Prosocial process because core design principle 4 (monitoring agreed behaviors) will push the group to think about how its activities in pursuit of a goal can be monitored. As with specificity, measurable goals afford selection criteria: if group members cannot say how they'll know that a goal has been accomplished, then success or failure cannot be clear. Wherever possible, it's usually helpful to include quantitative targets so that people know how many, what percentage, or what rate will count as an accomplishment. However, we recognize that in the complexity of human relations, it can sometimes be extremely difficult and time-consuming to collect the data needed to quantify whether a goal has been reached.

Consider, for example, the Canberra Innovation Network's goal of increasing collaboration between member organizations. How should that goal be assessed? Should it be measured by the number of meetings they have, the number of conflicts, the number of projects that are brought to shared completion, or perhaps the number of times they attend conferences together overseas? In most real-world situations, it's helpful to quantify goals as indicators wherever possible; however, at some point it may also be necessary to make a subjective judgment regarding progress. When it comes to measurement, the perfect should not be allowed to become the enemy of the good. What's important is the intent to collect reasonably objective data, rather than having a perfectly accurate and complete system of quantification.

Goals that are difficult and challenging on the one hand but attainable with effort on the other are more likely to lead to improved performance than goals that are either too easy or impossibly hard. By definition, improved performance requires group members to change behavior. If they can accomplish the goal too easily with minimal or even no change in behavior, then goal setting is beside the point. Conversely, members will think it unlikely that they can reach goals that are too difficult. If even major changes in behavior would leave them far from a goal, then it's unlikely that the motivational impact of possible goal achievement would take hold.[3]

Goals need to be relevant to individual and group learning and performance: people need to see that the goal fits within their domain of responsibility, and that it matters. For goals to work they need to be meaningful to group members and, ideally, inspiring. If goals are not consistent with collective and personal values, then groups are less likely to achieve them. When setting goals, members need to take into account the individual strengths of members rather than implement the goal in a "one size fits all" fashion. Relevant goals are worthwhile, and timely; they fit with the other efforts of the group and with the current environment, and the right people and resources are present to accomplish the goal with effort.

Small groups commonly make mistakes with this step. If an overall goal for the group is seen as relevant to only a limited number of group members, or to specific subgroups, then it won't motivate others to strive to attain the goal. A group can often avoid this problem by thinking through how subgoals in a variety of specific areas can support the overall goal. For example, if the goal is to increase group membership by 25 percent over the next year, but only one subgroup manages membership drives, a broad effort by the group is unlikely. But if the group can examine that goal more broadly, linking the subgoal to the broader goals of success for the group, other subgroups might be engaged in ways that would also foster increased membership. For example, the subgroup that manages the website might have a goal for increased traffic, including to pages that mention membership benefits. A subgroup that handles social media could try to increase tweets or Facebook posts, including those that mention how to become a member, and so on.

Finally, goals should be time-bound. That doesn't just mean that there should be a date of accomplishment; it means that there is a reasonable number of clear steps that members can envision happening over days, weeks, and months beginning immediately and extending forward until the goal has been accomplished. One benefit of time-bound goals is that normal day-to-day tasks are less likely to assume higher priority, draining away needed effort linked to longer-term goals.

The well-known acronym SMART summarizes goals that take into consideration the aspects we just discussed. A SMART goal is:

- Specific
- Measurable
- Attainable with effort
- Relevant
- Time-bound

Processes to Help Set SMART Goals

Many different approaches can work to help establish goals, but we like approaches that integrate the following steps that we outline.

First, gather together the goals and purposes that group members suggested for the matrix quadrants while working on the CDPs, and place them in raw form in front of the group. Experienced trainers will have been on the lookout, from the very beginning, for concrete goals that could be made SMART goals and been filing them away for this part of the process. To maintain the trust within the group, it's important to use the wording used by the group as much as possible.

If goals weren't generated during that work, a separate brainstorming phase may be required. We've had good results with the following approach, sometimes called "brain writing": After the facilitator shares the topic and ideas from the matrix with the team, each member writes down ideas individually and then shares them with the facilitator and the group. This reduces the bias toward early ideas and gives everyone a voice, essentially allowing ideas to be generated in a broad way across the whole group before discussion.

Once a set of ideas has been generated, the group can discuss each and begin to refine and prioritize them, fitting them into SMART form wherever possible and examining their costs and benefits, and whether the group has the necessary capability and resources. To act as guides for immediate action, goals should be consistent with the group's longer-term vision but be something that can be attained in the next quarter to year.

The exact process for refining and selecting goals depends, in part, on the culture of the group and how they have agreed to handle decision making. If the Prosocial session is being conducted face-to-face, prioritization can often be done with voting processes, such as by giving each group member a handful of sticky notes and allowing them to place their "votes" on goals that have been written on whiteboards or poster paper hung on the walls of the meeting room. Online tools (see http://www.prosocial.world/digital-tools) can be used similarly.

Once goals are prioritized they should be allocated to people, with parameters for accountability. While it is certainly possible for a small group to have accountability for completing a goal, it's often useful to hold just one person accountable for ensuring that the goal moves forward. Shared accountability can sometimes lead to nobody feeling any particular pressure to complete the goal. For that reason it's usually best to assign the responsibility for making sure goals are completed to people or small groups before ending the Prosocial session.

Creating a Review Process

For complex human systems in particular, setting goals is only the first step to achieving them. It is at least as important to revisit the goals regularly, using them as a guide to behavior and revising them as more is learned about the system. Groups need a process to review and update goals. Groups that have created SMART goals using the Prosocial process usually find value in continuing to use the matrix, and in reviewing the steps to implement the CDPs as members pursue the tasks required to meet their goals.

Setting accountability for achieving goals is likely to raise concerns that might reasonably be included in the bottom-left quadrant of the matrix. Now that the group has acquired the skills to be able to discuss these previously undiscussed fears and concerns, it's usually possible to work through them with openness and authenticity. The individual or small group that's accountable for ensuring the goal is executed might conduct an individual or collective matrix with the specific goal as the focus. For example, the process might begin with "What is important to us about achieving [this goal]?" followed by an exploration of the concerns the group members might have and so on.

We've noticed that successful Prosocial groups begin to adopt psychological flexibility and the language of the CDPs as part of group culture. This can be useful, because groups need to work continuously to maintain these processes. It's important, however, not to let group members start to see mere references to these processes as equivalents to the harder work of building a flexible and effective group culture. If a more systematic refresher is needed, the group can utilize annual group retreats, long-term strategic planning processes, board meetings, program-evaluation processes, or other processes meant for reflection and planning as times to revisit the Prosocial process in a more thorough way.

Evaluation and Research

The Prosocial process has a firm scientific foundation. Groups that are able to establish greater psychological flexibility at the level of individuals and the group, and that are able to implement the CDPs, can be assured that these steps are highly likely to foster effectiveness. What is not yet fully known is whether the steps in this book are sufficient to establish these processes. Thus every group that uses the Prosocial method is wise to evaluate its progress in terms of both processes and outcomes.

There are a set of assessments and group support tools available at http://www.prosocial.world. Groups that use the site can do so knowing that they are using evaluation tools that have been carefully designed to assess the Prosocial method. At the same time, by using the resources of the site, your experiences become part of a developmental process, such that you can expect to see increasingly refined support tools and assessment procedures.

Strategic Mapping with Government Agencies in Australia and Overseas

The following minicase explores how Robert Styles, an organizational consultant who has been instrumental in testing and extending Prosocial methods, created a clear sense of shared purpose and set goals with a number of government agencies in Australia, Southwest Asia, and Africa. In this instance, the mapping process was focused on shared purpose; however, the same approach could be used for implementing all of the core design principles. Robert shared his approach in an email to Paul:

> Generally, when I work with larger organizations with several divisions, or multiple organizations that are striving to work together to achieve a superordinate goal, I facilitate a series of workshops designed to clarify their shared purpose [of core design principle 1]. This involves a combination of divergent and convergent thinking. First those involved explore what a preferred and probable future could look like, particularly by considering how various system level trends and drivers will likely shape their behavior over time. Once this is done, together they benchmark a range of opportunities they can pursue that will begin taking them in that direction.

For each of Robert's three workshops, typically held about three or four weeks apart, he'll invite those involved to consider several questions, then lead them through four steps, in the following flow (reproduced with permission).

Workshop One: Envisioning

Pre Work

- What adaptive challenges is the organization facing now and in the future?
- What trends and drivers are impacting and shaping the behavior of the organization?

Session Work

1. Ask the group, "What's most important now for the long run?" Essentially this step involves completing the sentence stem, "What could it look like in 10 to 20 years if…"

2. Explore with the group which critical or uncertain trends and drivers will likely assert a bearing on their effort, and chart the adaptive challenges they will likely encounter. This step involves reviewing existing evidence and possibly conducting new research. The groups involved list all the stakeholders who would have a legitimate perspective on the strategic question being considered. All relevant perspectives are documented. Often, this step leads to a series of conversations with stakeholders.

3. Identify broad systemic trends and drivers and define "axes of uncertainty" that will delineate the future scenarios (e.g., climate change, large or small; increasing separatism between nations or not; the emergence of the gig economy; urbanization and the emerging power of "the city"). A meta-analysis of the stakeholders' views identifies the contingencies that will likely shape the behavior of the system over time, be they social, technological environmental, economic, political, or institutional.

4. Draft future scenarios based on the research—particularly preferred and probable scenarios that explore a portfolio of outcomes, including no-regret moves, real options, and big bets.

Workshop Two: Goal Setting

Pre Work

- What could your group's vision and goals be?
- Explore a portfolio of outcomes, including more or less ambitious and risky goals.

Session Work

1. Draft vision/goal based on your preferred and probable future that reflects a response to the trends and drivers.
2. Identify key criteria for success. List the positive impacts required to nudge those trends and drivers across the system in the right direction.
3. Identify opportunities and qualify how well/poorly they respond to the key criteria for success—the positive impacts required. These are real projects—often a combination of no-regret moves, real options, and big bets.
4. Rank each opportunity "fit for vision." Identify how each opportunity will explicitly nudge the trends and drivers in the right direction.

Workshop Three: Mapping to Capabilities

Pre Work

- What are your group's and organization's top capabilities and resources?
- What capabilities in the broader system most help your team achieve its goals?

Session Work

1. Assess capabilities required to achieve each opportunity against criteria from "gap" to "excellent." These are both hard and soft capabilities, everything from political will to having the right hammer.
2. Rank what is required for us to pursue each opportunity. This manages risk and shores up commitment. Some opportunities may be a great idea, but if the system lacks the capacity or investment to deliver, they may not be worth pursuing.
3. Prioritize opportunities based on "fitness for vision" and "fitness for capability."
4. Draft a road map underpinned by SMART actions over time to achieve the opportunities and vision.
5. Compare each group's road map and identify efficiencies, common elements, and actions.

Conclusion

Many organizations and groups have gone through goal-setting exercises before. Virtually every business or organizational consultant is, to a degree, familiar with these methods. Despite that, goal setting within the Prosocial process has a qualitatively different feel because it's a natural expression of the collective purpose of the group, done in a way that empowers its members to have a values-based journey and situated in the context of learning, adaptation, and flexibility engendered by the processes and ideas described throughout this book.

With the CDPs integrated, the group will realize that of course it makes sense to set goals that are specific, measurable, and time-bound—that's just core design principle 4 (monitoring agreed behaviors) in action. Of course the group will want to talk about who's responsible and what will happen when the goals are met (core design principles 2 and 5, equitable distribution of contributions and benefits and graduated responding to helpful and unhelpful behavior, respectively). It will feel natural to the group to get relevant people on board with the goal—that is, core design principle 3, fair and inclusive decision making. And if the group hits a rough spot as it pursues the goal, members will know what to do (core design principle 3).

Goal setting is vital. Every group knows that. The Prosocial process brings group vitality to this important process.

CONCLUSION

The More Beautiful World That Our Hearts Know Is Possible

In his book titled *The More Beautiful World Our Hearts Know Is Possible*, Charles Eisenstein asks: "Is it too much to ask, to live in a world where our human gifts go toward the benefit of all? Where our daily activities contribute to the healing of the biosphere and the well-being of other people?"[1] In the world of *Homo economicus*, this request has indeed become illegitimate, as the processes of top-down control and bottom-up markets have come to eclipse the processes of the commons. The human community faces very real problems. One only has to turn on the nearest screen to be aware of that. Worldwide, our politics are in turmoil, populations are shifting, and the economic order is changing. Science and expertise are under attack, authoritarianism is on the rise, and enormous practical problems such as global climate change and biodiversity loss are being addressed in ways that all thoughtful people admit are inadequate.

The physical sciences are not going to be enough on their own to protect humanity from the cataclysms we face. But, in our view, modern multidimensional and multilevel evolutionary science, combined with functional contextual-behavioral science, *do* appear to give us a way forward. In other words, we have argued in this volume that there is enough scientific information at hand to begin to undermine selfishness, greed, and apathy and to foster a cultural and spiritual transformation.

Building the Commons

Human beings cooperate vastly more than any other species because our symbolic abilities allow us to create shared fictions that can coordinate activity on any scale. For example, we have invented powerful tools of cultural organization and cooperation, such as laws, money, rights, and communication systems. Our stories and shared assumptions have lifted us up and carried us forward across the millennia, and there is no reason they cannot continue to do so, empowered by the guidance of scientific principles.

Unfortunately, these same symbolic systems can also push us down and hold us back when we allow ourselves to hold stories that are out of tune with the functional realities of human behavior and its biological and contextual determinants—most especially the myth of *Homo economicus*, which suggests that cooperation is merely a transaction and that somehow greed will lead to good. *Homo economicus* is a pseudoscientific falsehood that is ruining the planet, while at the same time creating an enormous backlog of inequity, loss of meaning, and loneliness. When we see humans as fundamentally and solely self-interested, then private markets and public regulation are the only two ways of organizing human cooperation at a large scale. Both are inherently limited tools for fostering human well-being. On the one hand, markets of privately owned goods and services support global trade and increased material wealth, but unrestrained they also support separation from others, exploitative relationships, unsustainable use and destruction of resources, and needless inequalities. On the other hand, while rules and regulations allow for global agreements and shared action in the public interest, hierarchies of power also ultimately rest upon coercion and thus create the conditions for loss of personal choice and meaning, insensitivity to the particular needs of individuals and communities, and disengagement from creative efforts to foster growth and prosperity. Humankind needs a middle path.

Elinor Ostrom won the Nobel Prize precisely because her work on the commons provided that hope. Let's briefly review some of the territory we have covered in this book. Small groups, and nested groups of groups, can create the conditions to sustainably coordinate their efforts to achieve a shared purpose. In this sense, the commons is always local; it is composed of particular people, in a particular situation, who care about achieving a particular shared purpose. Instead of being driven by the greed of markets or the fear of breaking rules, the commons is driven by shared caring for particular purposes and particular others.

The commons is by no means a universal panacea. Like markets and hierarchical regulation, the commons also has both strengths and weaknesses. While "commons thinking" can create sustainable, equitable, engaged, and purposeful small groups, it relies upon a kind of trust that is slow to build and easy to destroy. Indeed, the fragility of the commons initially led economists to talk about it as a tragedy because they forecast that *any* shared effort would ultimately be doomed to exploitation by selfish interests. Furthermore, because of its situatedness in a particular context, it has historically been difficult to imagine how the commons can be scaled to the levels of global cooperation required today.

We believe that one reason humans have struggled to imagine how the commons can be scaled is because we lacked an adequate understanding of multilevel and multidimensional evolution. As you have learned, a key tenet of multilevel selection theory is between-group competition (see chapter 1). Does this mean that humanity is forever doomed to engage in the sorts of competitive and destructive struggles we see between nations at war? Does it undermine the possibility of ever forming a more harmonious global society?

Our view is that it does not. Variation and selection mechanisms for groups themselves can be nested, as core design principle 8 suggests. For example, the Olympic Games are fantastically competitive in ways that have contributed to amazing increases in physical performance. At the same time, the Olympics require extraordinary cooperation at higher levels of organization simply to exist. Naked no-holds-barred competition in which teams could cheat and use methods to win outside of the established rules would destroy the Games if left unmonitored (core design principle 4) and allowed to persist without sanctions (core design principle 5), as modern doping scandals demonstrate. In a similar way, the kind of predatory competition that allows one business organization to thrive and another to go out of business due to practices that strain the bounds of decency not only creates undue suffering, it undermines capitalism itself. But like the Olympics, group selection processes can be managed in a way that fosters human progress.

Suppose various economic groups have been selected and retained through being more likely to survive harsh conditions (for example, a team surviving an economic downturn) or through enticing members of other groups to migrate because they see the group's advantages (for example, successfully hiring the members of other teams). If the practices that led to those outcomes are shared as part of a nested multilevel system (core design principle 8), all groups can benefit.

What's more, understanding that between-group competition can occur at symbolic and cultural levels allows for the possibility of even less-coercive forms of competition. A group can "win" over another simply by being so good at what it does that other groups wish to copy its practices. When selection between groups happens through such peaceful processes as migration from one group to another, or through the copying of cultural practices, it is the "loser" who is doing the deciding. The "vanquished" group isn't destroyed or forced to change: it is enabled. It's members change their practices because they see more and better opportunities.

Between-group competition doesn't necessarily mean between-group violence, nor does multilevel selection mean we're doomed to forever be locked in no-holds-barred struggles with peoples from other nations or cultures. Indeed, the core design principles (CDPs) and principles of psychological flexibility that are part of the Prosocial process suggest how we can use our knowledge of evolutionary principles to avoid that fate.

Perhaps another possible reason we've struggled to imagine how to scale the commons is the historical tendency of evolutionary and cultural scientists to observe, describe, and predict but not to change and influence. This tendency was, and still is, so dominant that testing ideas against the criterion of how useful they are for creating intentional change was almost unheard of in these fields of knowledge. Applied evolutionary science exists, but just barely, and this is especially so for the field of cultural change.

Part of that problem emerges out of mechanistic assumptions that can lull scientists into believing that once a system or domain is adequately modeled, it will be obvious how to create change—just pull the right levers! Unfortunately, it's hard to know if that's true because it's never clear when a system or domain is modeled "adequately"; thus, the test of application can be put off indefinitely without notable discomfort. This posture creates an anomaly in which application can be universally agreed to be important, but the responsibility for accomplishing that task is passed along to unseen others in an unknown future.

Even worse, as is shown in figure 16.1, mechanistic assumptions lead to the idea that there is an inverse relationship between basic knowledge and application. If application flows naturally out of a thoroughly adequate basic understanding, then dealing with the messy matter of real-world behavior change could actually *slow* scientific progress. High-quality basic research, this thinking goes, requires highly controlled environments, not the hurly-burly world of making a difference in existing groups.

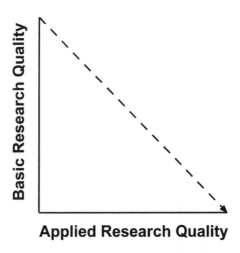

Figure 16.1. The typical mechanistic view of the relationship between basic and applied research quality.

We built this book using a different strategy and set of assumptions. In our contextual behavioral approach, the very purpose of psychological and cultural knowledge is understood to be the prediction *and influence* of the historically and contextually situated actions of people, individually and in groups. As is shown in figure 16.2, these contextualistic assumptions envision a mutually supportive relationship between basic knowledge and application because of that shared purpose. Said in another way, basic analyses should contain modifiable processes that can be used in application, while application tests the generality of those processes. This idea is characteristic of contextual behavioral science,[2] but parts of it exist in biological science too. Evolutionary biology, for example, is used to demanding field studies as an important aspect of a coherent research program. In our view, the Prosocial program is a kind of field station for cultural evolutionary studies.

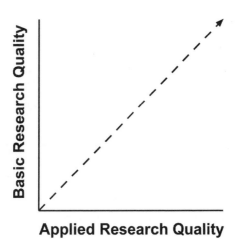

Figure 16.2. The functional contextual view of the relationship between basic and applied research quality.

The Prosocial program turned to psychological intervention science to help carry the CDPs forward as a coherent program of prosocial development and change. By combining the CDPs with the principles of psychological flexibility contained in acceptance and commitment training, a method of cultural change emerged built on psychosocial processes that are known to foster open and aware engagement in values-driven behavior change. The Prosocial process extends Ostrom's work to a practical method for balancing self and collective interest, while creating the conditions needed to construct trust and a shared commitment to thrive. We believe and hope we have shown that the combination of psychological flexibility processes and the CDPs can be extended to groups focused on any prosocial purpose, not just the protection of common-pool resources, and can guide trainers, organizers, and group members through the how-to steps needed to establish highly effective groups.

Our work is about both thinking globally and acting locally. Participants are uplifted not just by the toolkit, but by a new shared story about the nature of humanity—a story that provides realistic hope that we can do something about our challenges, locally and globally, by cultivating shared purpose while also managing the inevitability of self-interest in cooperative situations.

In this new story, human beings can develop and sustain shared commitments to improving the quality of our relationships, conversations, and agreements so as to ensure that self-interest is protected but cannot dominate collective interest. If we had to summarize in a sentence the journey we've taken in this book, we'd say it this way: *Complex adaptive systems of cooperation can grow out of more psychologically flexible people and environments, in which small groups, and groups of groups, create shared and principled agreements that establish trust and satisfy individual and group interests.* Or, perhaps more poetically, we can foster cooperation within and cooperation without, by restraining selfishness within and without, in the service of human values at all scales.

The World Our Hearts Imagine

Imagine for a moment your ideal world, restrained only by the finite reality of organisms that live and die as part of the natural world.

Actually do it. Allow your heart to imagine.

You might have in mind an image of a society in which everybody feels connected to one another in small networks of friends, family members, and colleagues; or perhaps your ideal world includes living in harmony with other elements of the natural world; or perhaps your ideal world is one without racism and hatred.

No matter what your unique vision contains, we are willing to bet that it includes harmonious and caring relationships with other people. We can be reasonably confident of that because being connected to other human beings in rich, meaningful, and caring ways is a primary human need based on multilevel selection and shared by all.[3]

If you actually imagined your ideal world, we wonder whether you *also* might have noticed a competing voice of skepticism, or even cynicism. Perhaps you noticed yourself having thoughts like *Nice exercise, but that will never happen*, or *There will always be people in power who look after themselves first*, or perhaps just some version of *I don't need to do that exercise. I've done it before and I know it leads nowhere.*

We see this tension in our work with groups all the time. People love to remember times when they were at their best, when they were most alive and vital. Groups generally love to talk about their visions for a better world, and the contributions their group might make to that. But at the same time, there is almost always someone in every group who expresses cynicism that the group could ever be more effective or make a bigger contribution to the world. We suspect these brave souls are expressive of a similar voice within all of us.

Any social vision contains the seeds of the same dialectic. Consider one of the most popular songs of all time: John Lennon's "Imagine." In it, Lennon paints a picture of a world of global cooperation, or at least global harmony. On the one hand, the song beautifully articulates our longing for a world of connection rather than separation. On the other hand, the song is also widely derided as being idealistic—the words of a hippie rather than a visionary.

We can think of this dynamic as a tension between love and fear. By "love," we don't mean romantic love but rather that force that draws us *toward* experiences and *toward* human cooperation. It is the approach element of human experience. By "fear," we mean that element of experience that causes us to withdraw—from our experiences and from each other. It is the *away* element of human experience.

As we explored in this book, in the very moment we imagine what we most care about, we're likely to feel vulnerable—to feel woundable. That's the nature of the human mind. Daring to care invariably contains within it regret over past actions that took us away from that which we care about, or the pain of potential failure, betrayal, or loss.

The more deeply we care, the more painful these tendrils of relating to the past, present, and future become. And before we even recognize it happening, we pull back. When caring for the world becomes a source of pain, we tighten into self-protective cynicism. Our disappointment and caring calcifies as a hardened, self-protective shell. There is a meme, usually attributed to George Carlin, that goes, "Scratch the surface of any cynic and you will find a disappointed idealist." If we don't imagine that a better world is possible, we cannot be disappointed. Stephen Colbert put it this way: "Cynicism masquerades as wisdom, but it is the farthest thing from it. Because cynics don't learn anything. Because cynicism is a self-imposed blindness, a rejection of the world because we are afraid it will hurt us or disappoint us."[4]

Worldwide the raw materials needed to justify our cynicism are in abundance right now. Trust in institutions such as the press and governments are at historic lows. Inequality is rapidly increasing as wealth has been concentrated with the super-rich more than at any time in the past 120 years,[5] often supported by tax-avoidance regimes that are legal but deeply unethical. These issues sit alongside catastrophic wars, widespread resistance to resettling refugees, widening gaps between rich and poor, exploitation of natural resources at the expense of people and other species, glacially slow responding to human-induced climate change, rates of depression spiraling upward, disengagement from work, loss of faith in leaders, and a weakening sense of community in many Western societies.

Contacting our hopes for a better world necessarily means contacting our pain over what has happened, what is happening, and what might happen in the future. It is easy to let our hearts harden over and let ourselves become cynical, as though to hope, wish, or *act* to create a more cooperative and harmonious world is the height of foolish naivety. Faced with all that pain, it is so easy to stop imagining the more beautiful world our hearts know is possible.

That dialectic is why we believe the combination of psychological flexibility and the CDPs is so important. If we are able to establish the CDPs in groups, we can look to a vast body of knowledge that suggests that these groups will be more able to act in the common interest. But trust is not easy, and the flip side of purpose is emotional vulnerability. Thus, as an applied matter it is not enough to know the ends we seek in group behavior;, we also need to provide the means to reach those ends. By giving group members a way to open up and care, at the level of the individual and of the small group, voluntary group members are provided creative ways of being and doing within the group that can rein in self-interest in the service of moving toward the values chosen by the group as a whole. This is not a guess, because we know that psychological flexibility fosters positive social relationships and commitments to them.[6]

This book is about articulating ideas for a scalable approach to creating greater cooperation globally, one group at a time. The toxic story of humanity as inherently selfish has harmed the human community for long enough. It is time to take back our own agency, and to create a story of humankind and the possibility of group action that fits better with the evolutionary dynamics of that social primate called a human.

Hope is the enemy of cynicism. The cynic brands the person with hope as "naive." To be naive is to be silly, stupid even. It takes real courage to open one's eyes and see the world's suffering and not retreat, but instead to remain open and hopeful. Stepping forward takes courage and motivation, but it also requires a change in how we even think about others. If we see ourselves as parts of a whole, then it's easier to step into the perspective of others. Put simply, when we realize that we are not separate, we see that we are part of historical processes of change. We can be less about "me" and more about "we." Nearly sixty years ago Martin Luther King Jr. expressed this idea this way:

> And all I'm saying is simply this: that all life is interrelated. We are tied in a single garment of destiny, caught in an inescapable network of mutuality. And what affects one directly affects all indirectly. As long as there is poverty in this world, no man can be totally rich, even if he has a billion dollars. As long as diseases are rampant and millions of people cannot expect to live more than twenty-eight years, no man can be totally healthy, even if he just got a clean bill of health from Mayo Clinic. Strangely enough, I can never be what I ought to be until you are what you ought to be, and you can never be what you ought to be until I am what I ought to be. This is the way the world is made.[7]

The Future of Prosocial

The authors of this book will not be the ones to evaluate whether the Prosocial process has a long-term impact or not. You and others like you will make that determination. You are the perfect agent for change. The fact that you have read this book suggests that you have prosocial preferences. And in that sense, you are particularly important. Experimental research in cooperation has shown that the evolution and stability of cooperation depends particularly on the ability of those with prosocial preferences to alter their social networks.[8]

Human culture is evolving increasingly rapidly. Who would have imagined even a few years ago that we would have systems with which people voluntarily share their houses with complete strangers; create new global currencies; publish their ideas cheaply, instantly, and globally; and create global communities of conversation? We are the ones who will choose the future we inhabit, and we need to apply our large brains to designing that future. We are delighted you have chosen to join us for the ride.

It takes only a single individual to begin. If you've read this far, that next step now passes to you.

Endnotes

Introduction
1. De Tocqueville (1835).
2. Travers & Milgram (1969).
3. Bhagat et al. (2016).
4. Greene (2014).
5. Kasser et al. (2007).

Chapter 1
1. Wilson (2019).
2. Darwin (1859), 490.
3. Wilson (2007).
4. Darwin (1871), 166.
5. Wilson & Wilson (2007), 345.
6. Examples of this research can be found in Muir (1985, 1996).
7. Sheldon, Sheldon, & Osbaldiston (2000).
8. Seeley (2010).
9. Hayes & Sanford (2014).

Chapter 2
1. Hardin (1968).
2. Hardin (1968), 1244.
3. Hardin's preferred solution is coercion through regulation. "The social arrangements that produce responsibility are arrangements that create coercion of some sort" (1968, 1247). He offers examples of how laws and punishments are the only way to stop people from robbing banks (thereby treating banks as a common-pool resource) or to encourage them to pay taxes.
4. Frank, Gilovich, & Regan (1993).
5. Common Cause Foundation (2016).
6. Ostrom & Parks (1973).
7. Ryan & Deci (2017).
8. Bollier (2016).
9. Bollier (2014), Kindle Edition. Location 300 of 3278.

10. Bollier (2014) Kindle Edition. Location 2543 of 3278, "The commons must be understood, then, as a verb as much as a noun."
11. LabGov (n.d.).
12. P2P Foundation (n.d.), case study 12.
13. https://primer.commonstransition.org/4-more/5-elements/case-studies/case-study-mutual-aid-networks.
14. Mutual Aid Networks (n.d.).
15. P2P Foundation (n.d.), case study 9.
16. P2P Foundation (n.d.), case study 1.
17. Loong Wong & Argumedo (2011).
18. P2P Foundation (n.d.), case study 8.
19. Enspiral (n.d.).
20. Bollier (2014), Location 244 of 3278.

Chapter 3

1. Covey (2012), 20.
2. Ryan & Deci (2017).

Chapter 4

1. We are deliberately not addressing sensitization in this chapter because it raises matters that are too technical for our purposes, but readers should know that at times repeated stimulation can lead to *greater* responsivity, especially when the stimuli are of a type that can harm the organism. We do not actually know how old habituation is, but there is clear evidence of habituation in slime molds, which evolved 600 million years ago (see Boisseau, Vogel, & Dussutour, 2016).
2. The ability to learn based on contingencies seems to be part of why so many new species quickly arrived during the Cambrian Period, the so-called Cambrian explosion. For one thing, organisms could now seek out specific environments, dramatically changing the selection pressures for particular phenotypic adaptations. For example, a bird whose behavior of digging into the riverbed was reinforced by finding small crustaceans could consistently seek out environments of that kind. Variations in beak structure could then be selected for based on their efficiency in finding and filtering out these crustaceans. The flamingo's odd beak is the result of such selection. This argument is laid out by Ginsburg & Jablonka (2010). A good book that explores the interaction of operant learning and other evolutionarily established processes is *The Science of Consequences* (Schneider, 2012).
3. This is an example of "stimulus equivalence." Although it is readily shown in human infants, after decades of trying it has not been reliably produced in nonhumans. A good initial review of the literature can be found in Hayes, Barnes-Holmes, & Roche (2001). Children who do not show stimulus equivalence do not develop normal human language (see Devany, Hayes, & Nelson, 1986).
4. As mentioned in an earlier chapter, the most likely source of the "two-way street" and combinatorial properties of human language is cooperation: a characteristic sound for an object resulted in others in the troop supplying the object when they heard the sound,

based on cues that the speaker wanted it. That analysis is laid out in detail in Hayes & Sanford (2014). We do not know how ancient symbolic relations of this kind are because we do not know whether hominins could make these relations. Chimpanzees do not (Dugdale & Lowe, 2000), so the ability postdates our common ancestors with chimpanzees.

5. One way to view meditation and mindfulness is as a rebalancing of this force, perhaps explaining why it is currently so popular.

6. There are many books on ACT, but see the second edition of the original book by Hayes, Strosahl, & Wilson (2011).

7. Dougher et al. (2007).

8. In relational frame theory (RFT), the technical term for this motivational effect is *augmenting*. Symbolic networks that form new consequences of importance are called *formative augmentals*, while those that increase the behavioral impact of existing consequences are called *motivative augmentals*. The original RFT book (Hayes, Barnes-Holmes, & Roche, 2001) is a good place to start if you wish to understand augmentals and the empirical support for them.

9. Wenzlaff & Wegner (2000).

10. Goubert et al. (2004).

11. See, for example, Bond, Lloyd, & Guenole (2013); Hayes et al. (1996).

12. A good summary of the data on defusion can be found in Hayes, Strosahl, & Wilson (2011).

13. For a recent and accessible review see Hayes (2019).

14. Van den Bos, van Dijk, & Crone (2011).

15. One corner of the literature on this can be found under the "social connectedness model," developed by Roger Vilardaga (for example, see Vilardaga et al., 2012, and Levin et al., 2016).

16. Parks, Joireman, & van Lange (2013).

17. Morrison et al. (2014).

18. Berghoff et al. (2012).

19. Gagnon, Dionne, & Pychyl (2016).

20. Parks, Joireman, & van Lange (2013).

21. Declerck, Boone, & Emonds (2013).

22. Ibid.

23. Balliet & van Lange (2013).

24. Van Lange et al. (2013).

25. Veage et al. (2014).

26. Levin et al. (2016).

Chapter 5

1. In behavioral language we distinguish between public behavior that all can see and private behavior that includes thinking and feeling. Because this language is not widely

understood, we find that "outside/inside" is usually an adequate substitute. We avoid labeling this dimension "behavior/mind" because, from a behavioral perspective, even private mind behavior such as thinking and feeling is behavior that's under the control of context.

2. In technical terms, the toward/away dimension reflects a symbolic parallel to the distinction between positive and negative reinforcement.
3. Skinner (1973).
4. Examples from lottery winners are from McNay (2012). Examples from cancer survivors are from Hawkes et al. (2013).
5. This question is inspired by a similar question used by Polk et al. (2016).
6. Our aim here is not to teach you how to facilitate groups. You can find great books devoted to that topic, such as Bowman (2008).
7. For other ideas about how to do this, see the discussion forum at https://community.prosocial.world, or take a look at Polk et al. (2016), or search online for matrix videos.
8. Don't write these up on the whiteboard, however, as they are not actual responses for the group.
9. Polk et al. (2016).
10. ACT literature is voluminous and widely available. A good place to start is with broadly focused ACT self-help books, such as those by Hayes (2005) or Harris (2008).

Chapter 6

1. For additional resources on introducing the individual matrix, see www.prosocial.world or https://contextualscience.org/act_matrix.
2. For examples of scenario planning see Chermack (2004) and Kahane (2012).

Chapter 7

1. Lambert et al. (2013).
2. Mathieu et al. (2015).
3. Ashforth & Mael (1989); Chang & Bordia (2001); De Cremer (2002); De Cremer & Stouten (2003); van Vugt & Hart (2004).
4. Stewart, Courtright, & Barrick (2012).
5. Listed in full at https://handbook.enspiral.com/agreements/people.html.
6. Walton & Cohen (2011).
7. De Cremer & van Vugt (1998); Kramer & Brewer (1984).
8. Parks, Joireman, & van Lange (2013).
9. Van Lange et al. (2013).
10. Bartlett (2016).
11. Burbano (2016).
12. Carpenter & Gong (2016).
13. Creswell et al. (2005).
14. Kilduff & Galinsky (2013).

15. Using what is known as diectic framing of time and person, respectively (Villatte, Villatte, & Hayes, 2015).
16. Grant (2008).
17. Harvard Business Review Staff (2014).
18. Green, Gino, & Staats (2017).
19. Grant (2007); Ryan & Deci (2017).
20. http://www.ey.com/gl/en/issues/ey-beacon-institute-the-business-case-for-purpose.
21. We suspect that this is a vast overestimate of the sort of shared purpose we have in mind. We suspect that many people may have said they were purpose driven if their group simply had a purpose statement. But having a purpose statement is not the same thing as living a purpose. In our experience, very few groups truly live their purpose.
22. Here are some helpful resources you might consult. In his book *Start with Why*, Simon Sinek uses the power of stories of success and vitality to help clarify what really matters for individuals and groups. Elsewhere, there are strategic planning methodologies freely available that encourage groups to envision possible futures and imagine how their work might impact those futures. You can find one example of such a method in an article from *Prosocial Magazine*, in which Robert Styles discusses how he integrates a method called "strategic roadmapping" with the Prosocial process (Wilson, 2017). We also discuss this approach in chapter 15.

Chapter 8

1. Anderson & Patterson (2008). Indeed, the absence of any visible benefits in acting fairly in many situations has led some to conclude that "virtue is its own reward" (Turillo et al., 2002).
2. Zhen & Yu (2016).
3. Ackermann, Fleiß, & Murphy (2016).
4. Debove, Baumard, & André (2016).
5. Barclay, Skarlicki, & Pugh (2005); Bobocel & Gosse (2015); Latham & Pinder (2005).
6. Aquino, Tripp, & Bies (2006); Fox, Spector, & Miles (2001); Greenberg (2002); Skarlicki & Folger (1997).
7. Brockner (2006).
8. Cropanzano et al. (2001); Walumbwa, Cropanzano, & Hartnell (2009).
9. Greenberg (2006).
10. Tyler & Blader (2013).
11. Van den Bos & Lind (2002).
12. Tyler & Degoey (1995).
13. Kazemi & Eek (2008).
14. Schäfer, Haun, & Tomasello (2015).
15. Kazemi & Eek (2008).
16. Another way into conversations regarding equity for more reflective groups is to consider and discuss basic ethical tools, such as the Golden Rule ("Do unto others as you would

have them do unto you") or John Rawls's "veil of ignorance" (essentially, make choices as if you might be the person most disadvantaged by the choice).

17. Blake et al. (2015).
18. Babcock et al. (1995).
19. Sawaoka, Hughes, & Ambady (2015).
20. For some ideas about ground rules to use with groups having difficult conversations, see http://www.prosocial.world.
21. You could start with the question "What matters most to us about equitable distribution of contributions and benefits?" But we've found this question is a bit academic for most groups.
22. Theft (Greenberg, 2002), retribution and revenge (Aquino, Tripp, & Bies, 2006), sabotage (Ambrose, Seabright, & Schminke, 2002), and retaliation (Skarlicki & Folger, 1997).
23. Epley, Caruso, & Bazerman (2006); Moore (2005).
24. Batson et al. (1995).
25. "Care labor" is a term used by Rich Bartlett (for example, see P2P Foundation, 2017). In management literature, this would be considered "organizational citizenship behavior."
26. Robertson (2015).
27. Interestingly, you might think that talking about power might be a useful way into conversations about equity. Who holds the power and why? There is some evidence that drawing attention to high-power individuals can exacerbate their tendency to behave unfairly, because power (in the sense of being in control of critical resources) tends to liberate people from paying attention to social and normative pressures, and it can lead people to be more egotistical (Blader & Chen, 2012).
28. Brockner (2006).
29. Teresa Amabile cited in Brockner (2006).
30. Bobocel & Gosse (2015).

Chapter 9

1. Ostrom (2008).
2. Slemp et al. (2018).
3. Ryan & Deci (2017).
4. Cronin et al. (2015).
5. This study (Cronin et al., 2015) replicates what is found in animal species. The strength of dominance hierarchies in animals varies. Species with strong hierarchies, such as chimpanzees, cooperate less than species with more relaxed, weaker hierarchies, such as cottontop tamarins.
6. Stouten & Tripp (2009).
7. Kugler & Bornstein (2013).
8. For example, https://www.industry.gov.au/sites/g/files/net3906/f/June%202018/document/pdf/the_performance_and_characteristics_of_australian_firms_with_employee_share_schemes.pdf.

9. Van Vugt et al. (2004).
10. Van Bunderen, Greer, & van Knippenberg (2017).
11. Ryan & Deci (2017).
12. Heifetz, Grashow, & Linsky (2009).
13. De Cremer & van Dijk (2005).
14. Harrell & Simpson (2016); Williams (2014).
15. Maner & Mead (2010); Williams (2014).
16. Useem (2017).
17. Useful resources include Frederic Laloux's *Reinventing Organizations* (http://www.reinventingorganizations.com), the Enspiral Handbook (https://handbook.enspiral.com), and this article by Gary Hamel (https://hbr.org/2011/12/first-lets-fire-all-the-managers).

    Some online articles that describe innovative decision making include those found at Ouishare (https://medium.com/ouishare-connecting-the-collaborative-economy), Percolab Droplets (https://medium.com/percolab-droplets), and Gini (https://blog.gini.net/).

    A key feature of both the advice process and consent-based decision making is that anyone in the group is empowered to make proposals. This encourages group members to take responsibility for the collective success of the group. To get a feel for what this can look like in a typical meeting using Holacracy, see https://medium.com/org-hacking/holacracys-integrative-decision-making-process-f750d4b82abc. See also https://www.holacracy.org/governance-meetings.

    For making decisions using Loomio, see https://blog.loomio.org/2015/09/10/10-tips-for-making-great-decisions-with-loomio/ as a starting point.
18. http://www.reinventingorganizationswiki.com/Decision_Making.
19. The consent-based decision-making process was the one we eventually proposed to the peer-reviewed ACT trainers. Everyone in the group deserved to be part of decision making, but everybody in the group was also a volunteer with a busy life. Giving everyone the power to block a decision ensures accountability, but this power also needs to be tempered to ensure responsible blocking. We adapted guidelines for blocking from the Enspiral Handbook (https://handbook.enspiral.com). Our example proposal, complete with our guidelines for blocking, is available at http://www.prosocial.world.
20. Varra et al. (2008).
21. Kernis & Goldman (2006); Ryan & Brown (2003).
22. Bond & Bunce (2003).
23. Meleady, Hopthrow, & Crisp (2013).
24. Gill (2018).
25. Kross et al. (2011).
26. Bernstein (2012).
27. Freeman (n.d.).
28. See, for example, Christensen and Grant (2016) for Australian examples, or Kasdan and Markman (2017) for an American example. For details of the initiative in Iceland see https://www.citizens.is/portfolio_page/my-neighbourhood.
29. Lukensmeyer (2007).

30. https://voteflux.org.

Chapter 10
1. McChrystal (2015), 161.
2. https://www.discourse.org/about.
3. Griskevicius, Tybur, & van den Bergh (2010).
4. Bateson et al. (2013); Bateson, Nettle, & Roberts (2006); Nettle et al. (2013).
5. Yang et al. (2018).
6. Hamlin (2015); Hamlin, Wynn, & Bloom (2007).
7. Bartlett (2016).
8. Oc, Bashshur, & Moore (2015).
9. Bracken & Rose (2011).
10. For a recent review, see Raffone & Srinivasan (2017).

Chapter 11
1. Van Lange et al. (2013).
2. Fehr & Gächter (2002); Kerr et al. (2009).
3. Gürerk, Irlenbusch, & Rockenbach (2006); Fehr & Gächter (2002).
4. Goltz (2013).
5. Irwin, Mulder, & Simpson (2014).
6. Goltz (2013).
7. Irwin, Mulder, & Simpson (2014).
8. Ryan & Deci (2017).
9. Van Lange et al. (2013).
10. That is, not just to leaders but also to support staff; not just to those who are constantly speaking but also to those who work quietly but helpfully in the background; not just to men but also to women; not just to ethnic majority members but also to ethnic minorities.
11. Bamberger & Levi (2009); Cacioppe (1999).
12. Van Lange et al. (2013).
13. Shahzad & Muller (2016).
14. Jentz (2007).
15. Daniels & Daniels (2004); Braksick (2007).

Chapter 12
1. De Wit, Greer, & Jehn (2012).
2. Simons & Peterson (2000).
3. Tekleab, Quigley, & Tesluk (2009).
4. Simons & Peterson (2000).

5. Edmondson & Roloff (2009).
6. Bradley et al. (2012).
7. Langfred (2007).
8. Fisher, Ury, & Patton (1981).
9. Donald et al. (2019); Heppner et al. (2008).
10. Lakey et al. (2008).
11. Ury (2015).
12. Heifetz, Grashow, & Linsky (2009).
13. Guinote (2017).

Chapter 13

1. See http://www.cityrepair.org. This and other place-making projects are discussed from the perspective of the core design principles in Wilson, Marshall, & Iserhott (2011).
2. Ostrom (2003), 61.
3. https://en.oxforddictionaries.com/definition/subsidiarity.
4. Wilson, Kauffman Jr., & Purdy (2011).
5. Wilson (2017).
6. See https://www.buurtzorg.com/about-us/buurtzorgmodel. Laloux (2014) provides an excellent description of Buurtzorg and a number of other organizations that score high on core design principle 7.
7. Cited in Laloux (2014).
8. Matta & Morgan (2011); Schaffer & Ashkenas (2005).
9. McChrystal (2015), 1.
10. Ibid, 96.

Chapter 14

1. Wilson & Wilson (2007); Wilson (2015).
2. Yong (2016).
3. Turchin (2016).
4. McGinnis (1999); Ostrom (2010b).
5. Ostrom & Parks (1973).
6. Boettke, Lemke, & Palagashvili (2016).
7. Wilson & Hessen (2014).
8. Aligica & Tarko (2012).
9. Jackall (2009).
10. Haidt & Wilson (2013).
11. See https://www.prosocial.world/magazine. The two quotes that follow are also taken from this interview.
12. New Belgium Brewing (2018).

13. Chen & Kelly (2014).
14. http://businessforpeace.no.
15. Saxegaard (2018).
16. Camus (1948). Kindle edition location 1486 of 3492.
17. This quote is from an interview that is online at https://evolution-institute.org/darwinian-business-conference-at-nyu-stern-school/.

Chapter 15

1. A good summary is provided by Heslin, Carson, & VandeWalle (2009).
2. Locke & Latham (2002).
3. Locke & Latham (1990).

Conclusion

1. Eisenstein (2013), 12.
2. Zettle et al. (2016).
3. Ryan & Deci (2017).
4. Colbert (2006). Commencement address at http://departments.knox.edu/newsarchive/news_events/2006/x12547.html
5. Neate (2017).
6. See, for example, Lenger, Gordon, & Nguyen (2017). Similar findings have been shown in a parental context, such as by Brockman et al. (2016).
7. King (1961). From Mieder (2010).
8. For an experimental demonstration, see Melamed, Simpson, & Harrell (2017).

References

Ackermann, K. A., Fleiß, J., & Murphy, R. O. (2016). Reciprocity as an individual difference. *Journal of Conflict Resolution, 60*(2), 340–367.

Aligica, P. D., & Tarko, V. (2012). Polycentricity: From Polanyi to Ostrom, and beyond. *Governance, 25*(2), 237–262.

Ambrose, M. L., Seabright, M. A., & Schminke, M. J. (2002). Sabotage in the workplace: The role of organizational injustice. *Organizational Behavior and Human Decision Processes, 89*(1), 947–965.

Anderson, W. D., & Patterson, M. L. (2008). Effects of social value orientations on fairness judgments. *Journal of Social Psychology, 148*(2), 223–245.

Aquino, K., Tripp, T. M., & Bies, R. J. (2006). Getting even or moving on? Power, procedural justice, and types of offense as predictors of revenge, forgiveness, reconciliation, and avoidance in organizations. *Journal of Applied Psychology, 91*(3), 653–668.

Ashforth, B. E., & Mael, F. (1989). Social identity theory and the organization. *Academy of Management Review, 14*(January), 20–39.

Aubé, C., & Rousseau, V. (2005). Team goal commitment and team effectiveness: The role of task interdependence and supportive behaviors. *Group Dynamics: Theory, Research, and Practice, 9*(3), 189–204.

Babcock, L., Loewenstein, G., Issacharoff, S., & Camerer, C. (1995). Biased judgments of fairness in bargaining. *American Economic Review, 85*(5), 1337–1343.

Balliet, D., & van Lange, P. A. (2013). Trust, conflict, and cooperation: A meta-analysis. *Psychological Bulletin, 139*(5), 1090–1112.

Bamberger, P. A., & Levi, R. (2009). Team-based reward allocation structures and the helping behaviors of outcome-interdependent team members. *Journal of Managerial Psychology, 24*(4), 300–327.

Barclay, L. J., Skarlicki, D. P., & Pugh, S. D. (2005). Exploring the role of emotions in injustice perceptions and retaliation. *Journal of Applied Psychology, 90*(4), 629–643.

Bartlett, R. D. (2016, May 19). Bootstrapping a bossless organisation in 3 easy steps ;-). Retrieved from https://medium.com/enspiral-tales/bootstrapping-a-bossless-organisation-in-3-easy-steps-afc653e8f5e6#.rcmbwfphu

Bateson, M., Callow, L., Holmes, J. R., Redmond Roche, M. L., & Nettle, D. (2013). Do images of "watching eyes" induce behaviour that is more pro-social or more normative? A field experiment on littering. *PLoS One, 8*(12), e82055.

Bateson, M., Nettle, D., & Roberts, G. (2006). Cues of being watched enhance cooperation in a real-world setting. *Biology Letters, 2*(3), 412–414.

Batson, C. D., Klein, T. R., Highberger, L., & Shaw, L. L. (1995). Immorality from empathy-induced altruism: When compassion and justice conflict. *Journal of Personality and Social Psychology, 68*(6), 1042–1054.

Berghoff, C. R., Pomerantz, A. M., Pettibone, J. C., Segrist, D. J., & Bedwell, D. R. (2012). The relationship between experiential avoidance and impulsiveness in a nonclinical sample. *Behaviour Change, 29*(1), 25–35.

Bernstein, E. (2012, February 7). Speaking up is hard to do: Researchers explain why. *Wall Street Journal*. Retrieved from https://www.wsj.com/articles/SB10001424052970204136404577207020525853492

Bhagat, S., Burke, M., Duik, C., Filiz, I. O., & Edunov, S. (2016, February 16). Three and a half degrees of separation. Retrieved from https://research.fb.com/three-and-a-half-degrees-of-separation

Blader, S. L., & Chen, Y. R. (2012). Differentiating the effects of status and power: A justice perspective. *Journal of Personality and Social Psychology, 102*(5), 994–1014.

Blake, P. R., McAuliffe, K., Corbit, J., Callaghan, T. C., Barry, O., Bowie, A., et al. (2015). The ontogeny of fairness in seven societies. *Nature, 528*(7581), 258–261.

Bobocel, D. R., & Gosse, L. (2015). Procedural justice: A historical review and critical analysis. In R. S. Cropanzano & M. L. Ambrose (Eds.), *The Oxford handbook of justice in the workplace* (Vol. 1, pp. 51–88). New York: Oxford University Press.

Boettke, P. J., Lemke, J. S., & Palagashvili, L. (2016). Re-evaluating community policing in a polycentric system. *Journal of Institutional Economics, 12*(2), 305–325.

Boisseau, R. P., Vogel, D., & Dussutour, A. (2016). Habituation in non-neural organisms: Evidence from slime moulds. *Proceeding of the Royal Society B: Biological Sciences, 283*(1829).

Bollier, D. (2016, June 25). Beyond development: The commons as a new/old paradigm of human flourishing [Blog post]. Retrieved from http://www.bollier.org/blog/beyond-development-commons-newold-paradigm-human-flourishing

Bollier, D. (2014). *Think like a commoner: A short introduction to the life of the commons*. Gabriola Island, BC: New Society Publishers. Kindle Edition.

Bond, F. W., & Bunce, D. (2003). The role of acceptance and job control in mental health, job satisfaction, and work performance. *Journal of Applied Psychology, 88*(6), 1057–1067.

Bond, F. W., Lloyd, J., & Guenole, N. (2013). The work-related acceptance and action questionnaire (WAAQ): Initial psychometric findings and their implications for measuring psychological flexibility in specific contexts. *Journal of Occupational and Organizational Psychology, 86*(3), 331–347.

Bowman, S. L. (2008). *Training from the back of the room!: 65 ways to step aside and let them learn*. Hoboken, NJ: John Wiley & Sons.

Bracken, D. W., & Rose, D. S. (2011). When does 360-degree feedback create behavior change? And how would we know it when it does? *Journal of Business and Psychology, 26*(2), 183–192.

Bradley, B. H., Postlethwaite, B. E., Klotz, A. C., Hamdani, M. R., & Brown, K. G. (2012). Reaping the benefits of task conflict in teams: The critical role of team psychological safety climate. *Journal of Applied Psychology, 97*(1), 151–158.

Braksick, L. W. (2007). *Unlock behavior, unleash profits: Developing leadership behavior that drives profitability in your organization*. New York: McGraw Hill.

Brockman, C., Snyder, J., Gewirtz, A., Gird, S. R., Quattlebaum, J., Schmidt, N., et al. (2016). Relationship of service members' deployment trauma, PTSD symptoms, and experiential avoidance to postdeployment family reengagement. *Journal of Family Psychology, 30*(1), 52–62.

Brockner, J. (2006, March). Why it's so hard to be fair. *Harvard Business Review*. Retrieved from https://hbr.org/2006/03/why-its-so-hard-to-be-fair

Burbano, V. C. (2016). Social responsibility messages and worker wage requirements: Field experimental evidence from online labor marketplaces. *Organization Science, 27*(4), 1010–1028.

Cacioppe, R. L. (1999). Using team—individual reward and recognition strategies to drive organizational success. *Leadership & Organization Development Journal, 20*(6), 322–331.

Camus, A. (1948). *The plague.* New York: Vintage.

Carpenter, J., & Gong, E. (2016). Motivating agents: How much does the mission matter? *Journal of Labor Economics, 34*(1), 211–236.

Chang, A., & Bordia, P. (2001). A multidimensional approach to the group cohesion-group performance relationship. *Small Group Research, 32*(4), 379–405.

Chen, X., & Kelly, T. F. (2014). B-Corps—A growing form of social enterprise: Tracing their progress and assessing their performance. *Journal of Leadership & Organizational Studies, 22*(1), 102–114.

Chermack, T. J. (2004). Improving decision-making with scenario planning. *Futures, 36*(3), 295–309.

Christensen, H. E., & Grant, B. (2016). Participatory budgeting in australian local government: An initial assessment and critical issues. *Australian Journal of Public Administration, 75*(4), 457–475.

Common Cause Foundation. (2016). *Perceptions matter: The Common Cause UK values survey.* Retrieved from https://valuesandframes.org/resources/CCF_survey_perceptions_matter_full_report.pdf

Covey, S. R. (2012). *The wisdom and teachings of Stephen R. Covey.* New York: Simon and Schuster.

Creswell, J. D., Welch, W. T., Taylor, S. E., Sherman, D. K., Gruenewald, T. L., & Mann, T. (2005). Affirmation of personal values buffers neuroendocrine and psychological stress responses. *Psychological Science, 16*(11), 846–851.

Cronin, K. A., Acheson, D. J., Hernández, P., & Sánchez, A. (2015). Hierarchy is detrimental for human cooperation. *Scientific Reports, 5*(18634). http://www.nature.com/articles/srep18634#supplementary-information

Cropanzano, R., Byrne, Z. S., Bobocel, D. R., & Rupp, D. E. (2001). Moral virtues, fairness heuristics, social entities, and other denizens of organizational justice. *Journal of Vocational Behavior, 58*(2), 164–209.

Daniels, A. C., & Daniels, J. E. (2004). *Performance management: changing behavior that drives organizational effectiveness* (4th ed.). Atlanta, GA: Performance Management Publications.

Darwin, C. (1859). *The origin of species.* London: John Murray.

Darwin, C. (1871). *The descent of man.* London: Murray.

Debove, S., Baumard, N., & André, J.-B. (2016). Models of the evolution of fairness in the ultimatum game: A review and classification. *Evolution and Human Behavior, 37*(3), 245–254.

Declerck, C. H., Boone, C., & Emonds, G. (2013). When do people cooperate? The neuroeconomics of prosocial decision making. *Brain and Cognition, 81*(1), 95–117.

De Cremer, D. (2002). Respect and cooperation in social dilemmas: The importance of feeling included. *Personality and Social Psychology Bulletin, 28*(10), 1335–1341.

De Cremer, D., & Stouten, J. (2003). When do people find cooperation most justified? The effect of trust and self–other merging in social dilemmas. *Social Justice Research, 16*(1), 41–52.

De Cremer, D., & van Dijk, E. (2005). When and why leaders put themselves first: Leader behaviour in resource allocations as a function of feeling entitled. *European Journal of Social Psychology, 35*(4), 553–563.

De Cremer, D., & van Vugt, M. (1998). Collective identity and cooperation in a public goods dilemma: A matter of trust or self-efficacy. *Current Research in Social Psychology, 3*(1), 1–11.

De Tocqueville, A. (1835). *Democracy in America.* New York: Penguin Classic.

Devany, J. M., Hayes, S. C., & Nelson, R. O. (1986). Equivalence class formation in language-able and language-disabled children. *Journal of the Experimental Analysis of Behavior, 46*(3), 243–257.

De Wit, F. R., Greer, L. L., & Jehn, K. A. (2012). The paradox of intragroup conflict: A meta-analysis. *Journal of Applied Psychology, 97*(2), 360–390.

Donald, J. N., Sahdra, B. K., Van Zanden, B., Duineveld, J. J., Atkins, P. W. B., Marshall, S. L., & Ciarrochi, J. (2019). Does your mindfulness benefit others? A systematic review and meta-analysis of the link between mindfulness and prosocial behaviour. *British Journal of Psychology, 110*(1), 101–125.

Dougher, M. J., Hamilton, D. A., Fink, B. C., & Harrington, J. (2007). Transformation of the discriminative and eliciting functions of generalized relational stimuli. *Journal of the Experimental Analysis of Behavior, 88*(2), 179–197.

Dugdale, N., & Lowe, C. F. (2000). Testing for symmetry in the conditional discriminations of language-trained chimpanzees. *Journal of the Experimental Analysis of Behavior, 73*(1), 5–22.

Edmondson, A. C., Roloff, K. S. (2009). Overcoming barriers to collaboration: Psychological safety and learning in diverse teams. In E. Salas, G. F. Goodwin, & C. S. Burke (Eds.), *Team effectiveness in complex organizations: Cross-disciplinary perspectives and approaches* (pp. 183–208). New York: Routledge.

Eisenstein, C. (2013). *The more beautiful world our hearts know is possible.* Berkeley, CA: North Atlantic Books.

Enspiral. (n.d.). Enspiral handbook. Retrieved from https://handbook.enspiral.com

Epley, N., Caruso, E. M., & Bazerman, M. H. (2006). When perspective taking increases taking: Reactive egoism in social interaction. *Journal of Personality and Social Psychology, 91*(5), 872–889.

Fehr, E., & Gächter, S. (2002). Altruistic punishment in humans. *Nature, 415*(6868), 137–140.

Fisher, R., Ury, W., & Patton, B. (1981). *Getting to yes: Negotiating agreement without giving in.* New York: Penguin Books.

Fox, S., Spector, P. E., & Miles, D. (2001). Counterproductive work behavior (CWB) in response to job stressors and organizational justice: Some mediator and moderator tests for autonomy and emotions. *Journal of Vocational Behavior, 59*(3), 291–309.

Frank, R. H., Gilovich, T., & Regan, D. T. (1993). Does studying economics inhibit cooperation? *Journal of Economic Perspectives, 7*(2), 159–171.

Freeman, J. (n.d.). The tyranny of structurelessness. Retrieved from http://www.jofreeman.com/joreen/tyranny.htm

Gagnon, J., Dionne, F., & Pychyl, T. A. (2016). Committed action: An initial study on its association to procrastination in academic settings. *Journal of Contextual Behavioral Science, 5*(2), 97–102.

Gill, L. (2018, March 22). 10 components that successfully abolished hierarchy (in 70+ companies) [Blog entry]. Retrieved from https://corporate-rebels.com/10-components-k2k

Ginsburg, S., & Jablonka, E. (2010). The evolution of associative learning: A factor in the Cambrian explosion. *Journal of Theoretical Biology, 266*(1), 11–20.

Goltz, S. M. (2013). A behavior analysis of individuals' use of the fairness heuristic when interacting with groups and organizations. *Journal of Organizational Behavior Management, 33*(1), 5–30.

Goubert, L., Crombez, G., Eccleston, C., & Devulder, J. (2004). Distraction from chronic pain during a pain-inducing activity is associated with greater post-activity pain. *Pain, 110*(1–2), 220–227.

Grant, A. M. (2007). Relational job design and the motivation to make a prosocial difference. *Academy of Management Review, 32*(2), 393–417.

Grant, A. M. (2008). Does intrinsic motivation fuel the prosocial fire? Motivational synergy in predicting persistence, performance, and productivity. *Journal of Applied Psychology, 93*(1), 48–58.

Green, P., Gino, F., & Staats, B. R. (2017). Seeking to belong: How the words of internal and external beneficiaries influence performance. Working paper 17-073. Retrieved from https://papers.ssrn.com/sol3/papers.cfm?abstract_id=2912271

Greenberg, J. (2002). Who stole the money, and when? Individual and situational determinants of employee theft. *Organizational Behavior and Human Decision Processes, 89*(1), 985–1003.

Greenberg, J. (2006). Losing sleep over organizational injustice: Attenuating insomniac reactions to underpayment inequity with supervisory training in interactional justice. *Journal of Applied Psychology, 91*(1), 58–69.

Greene, J. (2014). *Moral tribes: Emotion, reason and the gap between us and them.* London: Atlantic Books.

Griskevicius, V., Tybur, J. M., & van den Bergh, B. (2010). Going green to be seen: Status, reputation, and conspicuous conservation. *Journal of Personality and Social Psychology, 98*(3), 392–404.

Guinote, A. (2017). How power affects people: Activating, wanting, and goal seeking. *Annual Review of Psychology, 68*, 353–381.

Gürerk, Ö., Irlenbusch, B., & Rockenbach, B. (2006). The Competitive Advantage of Sanctioning Institutions. *Science, 312*(5770), 108. doi:10.1126/science.1123633

Haidt, J., & Wilson, D. S. (2013, September 5). Sears ignores the invisible band. *Forbes.* Retrieved from https://www.forbes.com/sites/darwinsbusiness/2013/09/05/sears-ignores-the-invisible-band/#3b4d024b1454

Hamlin, J. K. (2015). The case for social evaluation in preverbal infants: Gazing toward one's goal drives infants' preferences for Helpers over Hinderers in the hill paradigm. *Frontiers in Psychology, 5.*

Hamlin, J. K., Wynn, K., & Bloom, P. (2007). Social evaluation by preverbal infants. *Nature, 450*(7169), 557–559. doi:http://www.nature.com/nature/journal/v450/n7169/suppinfo/nature06288_S1.html

Hardin, G. (1968). The tragedy of the commons. *Science, 162*(3859), 1243–1248.

Harrell, A., & Simpson, B. (2016). The dynamics of prosocial leadership: Power and influence in collective action groups. *Social Forces, 94*(3), 1283–1308.

Harris, R. (2008). *The happiness trap: How to stop struggling and start living.* Boston: Trumpeter Books.

Harvard Business Review Staff. (2014, November). Cooks make tastier food when they can see their customers. *Harvard Business Review*. Retrieved from https://hbr.org/2014/11/cooks-make-tastier-food-when-they-can-see-their-customers

Hawkes, A. L., Chambers, S. K., Pakenham, K. I., Patrao, T. A., Baade, P. D., Lynch, B. M., et al. (2013). Effects of a telephone-delivered multiple health behavior change intervention (CanChange) on health and behavioral outcomes in survivors of colorectal cancer: A randomized controlled trial. *Journal of Clinical Oncology, 31*(18), 2313–2321.

Hayes, S. C. (2005). *Get out of your mind and into your life: The new acceptance and commitment therapy*. Oakland, CA: New Harbinger Publications.

Hayes, S. C. (2019). *A liberated mind: How to pivot toward what matters*. New York: Penguin/Avery.

Hayes, S. C., Barnes-Holmes, D., & Roche, B. (2001). *Relational frame theory: A post-Skinnerian account of human language and cognition*. New York: Kluwer Academic/Plenum Publishers.

Hayes, S. C., & Sanford, B. T. (2014). Cooperation came first: Evolution and human cognition. *Journal of the Experimental Analysis of Behavior, 101*(1), 112–129.

Hayes, S. C., Strosahl, K. D., & Wilson, K. G. (2011). *Acceptance and commitment therapy: The process and practice of mindful change* (2nd ed.). New York: Guilford Press.

Hayes, S. C., Wilson, K. G., Gifford, E. V., Follette, V. M., & Strosahl, K. (1996). Experimental avoidance and behavioral disorders: A functional dimensional approach to diagnosis and treatment. *Journal of Consulting and Clinical Psychology, 64*(6), 1152–1168.

Heifetz, R. A., Grashow, A., & Linsky, M. (2009). *The practice of adaptive leadership: Tools and tactics for changing your organization and the world*. Boston: Harvard Business Press.

Heppner, W. L., Kernis, M. H., Lakey, C. E., Campbell, W., Goldman, B. M., Davis, P. J., & Cascio, E. V. (2008). Mindfulness as a means of reducing aggressive behavior: Dispositional and situational evidence. *Aggressive Behavior, 34*(5), 486–496.

Heslin, P. A., Carson, J. B., & VandeWalle, D. (2009). Practical applications of goal-setting theory to performance management. In J. W. Smither (Ed.), *Performance management: Putting research into action* (pp. 89–114). San Francisco, CA: Jossey-Bass.

Irwin, K., Mulder, L., & Simpson, B. (2014). The detrimental effects of sanctions on intragroup trust: Comparing punishments and rewards. *Social Psychology Quarterly, 77*(3), 253–272.

Jackall, R. (2009). *Moral Mazes: The world of corporate managers*. New York: Oxford University Press.

Jentz, B. (2007). *Talk sense: Communicating to lead and learn*. Acton, MA: Research for Better Teaching, Inc.

Kahane, A. M. (2012). *Transformative scenario planning: Working together to change the future*. San Francisco: Berrett-Koehler Publishers.

Kasdan, A., & Markman, E. (2017). Participatory budgeting and community-based research: Principles, practices, and implications for impact validity. *New Political Science, 39*(1), 143–155.

Kasser, T., Cohn, S., Kanner, A. D., & Ryan, R. M. (2007). Some costs of American corporate capitalism: A psychological exploration of value and goal conflicts. *Psychological Inquiry, 18*(1), 1–22.

Kazemi, A., & Eek, D. (2008). Promoting cooperation in social dilemmas via fairness norms and group goals. In A. Biel, D. Eek, E. Gärling, & M. Gustafsson, *New issues and paradigms in research on social dilemmas* (pp. 72–92). Boston: Springer.

Kernis, M. H., & Goldman, B. M. (2006). Assessing stability of self-esteem and contingent self-esteem. In M. H. Kernis (Ed.), *Self-esteem issues and answers: A sourcebook of current perspectives* (pp. 77–85). New York: Psychology Press.

Kerr, N. L., Rumble, A. C., Park, E. S., Ouwerkerk, J. W., Parks, C. D., Gallucci, M., et al. (2009). "How many bad apples does it take to spoil the whole barrel?": Social exclusion and toleration for bad apples. *Journal of Experimental Social Psychology, 45*(4), 603–613.

Kilduff, G. J., & Galinsky, A. D. (2013). From the ephemeral to the enduring: How approach-oriented mindsets lead to greater status. *Journal of Personality and Social Psychology, 105*(5), 816–831.

Kramer, R. M., & Brewer, M. B. (1984). Effects of group identity on resource use in a simulated commons dilemma. *Journal of Personality and Social Psychology, 46*(5), 1044–1057.

Kross, E., Berman, M. G., Mischel, W., Smith, E. E., & Wager, T. D. (2011). Social rejection shares somatosensory representations with physical pain. *Proceedings of the National Academy of Sciences, 108*(15), 6270–6275.

Kugler, T., & Bornstein, G. (2013). Social dilemmas between individuals and groups. *Organizational Behavior and Human Decision Processes, 120*(2), 191–205.

LabGov. (n.d.). Bologna lab. Retrieved from http://labgov.city/explore-by-lab/bolognalab

Lakey, C. E., Kernis, M. H., Heppner, W. L., & Lance, C. E. (2008). Individual differences in authenticity and mindfulness as predictors of verbal defensiveness. *Journal of Research in Personality, 42*(1), 230–238.

Laloux, F. (2014). *Reinventing organizations*. Millis, MA: Nelson Parker.

Lambert, N. M., Stillman, T. F., Hicks, J. A., Kamble, S., Baumeister, R. F., & Fincham, F. D. (2013). To belong is to matter: Sense of belonging enhances meaning in life. *Personality and Social Psychology Bulletin, 39*(11), 1418–1427.

Langfred, C. W. (2007). The downside of self-management: A longitudinal study of the effects of conflict on trust, autonomy, and task interdependence in self-managing teams. *Academy of Management Journal, 50*(4), 885–900.

Latham, G. P., & Pinder, C. C. (2005). Work motivation theory and research at the dawn of the twenty-first century. *Annual Review of Psychology, 56*, 485–516.

Lenger, K. A., Gordon, C. L., & Nguyen, S. P. (2017). Intra-individual and cross-partner associations between the five facets of mindfulness and relationship satisfaction. *Mindfulness, 8*(1), 171–180.

Levin, M. E., Luoma, J. B., Vilardaga, R., Lillis, J., Nobles, R., & Hayes, S. C. (2016). Examining the role of psychological inflexibility, perspective taking, and empathic concern in generalized prejudice. *Journal of Applied Social Psychology, 46*(3), 180–191.

Locke, E. A., & Latham, G. P. (1990). *A theory of goal setting and task performance*. Englewood Cliffs, NJ: Prentice Hall.

Locke, E. A., & Latham, G. P. (2002). Building a practically useful theory of goal setting and task motivation: A 35-year odyssey. *American Psychologist, 57*(9), 705–717.

Loong Wong, B. Y., & Argumedo, A. (2011, July 4). The thriving biodiversity of Peru's Potato Park. Retrieved from https://ourworld.unu.edu/en/the-thriving-biodiversity-of-peru-potato-park

Lukensmeyer, C. J. (2007). Large-scale citizen engagement and the rebuilding of New Orleans: A case study. *National Civic Review, 96*(3), 3–15.

Maner, J. K., & Mead, N. L. (2010). The essential tension between leadership and power: When leaders sacrifice group goals for the sake of self-interest. *Journal of Personality and Social Psychology, 99*(3), 482–497.

Mathieu, J. E., Kukenberger, M. R., D'Innocenzo, L., & Reilly, G. (2015). Modeling reciprocal team cohesion-performance relationships, as impacted by shared leadership and members' competence. *Journal of Applied Psychology, 100*(3), 713–734.

Matta, N., & Morgan, P. (2011). Local empowerment through rapid results. *Stanford Social Innovation Review,* (Summer), 51–55.

McChrystal, S. (2015). *Team of teams: New rules of engagement for a complex world.* New York: Penguin.

McGinnis, M. D. (1999). *Polycentric governance and development: Readings from the workshop in political theory and policy analysis.* Ann Arbor, MI: University of Michigan Press.

McNay, D. (2012). *Life lessons from the lottery: Protecting your money in a scary world.* Lexington, KY: RRP International.

Melamed, D., Simpson, B., & Harrell, A. (2017). Prosocial orientation alters network dynamics and fosters cooperation. *Scientific Reports, 7*(357).

Meleady, R., Hopthrow, T., & Crisp, R. J. (2013). Simulating social dilemmas: Promoting cooperative behavior through imagined group discussion. *Journal of Personality and Social Psychology, 104*(5), 839–853.

Mieder, W. M. (2010). *"Making a Way Out of No Way": Martin Luther King's Sermonic Proverbial Rhetoric.* New York: Peter Lang Publishing.

Moore, D. A. (2005). Myopic biases in strategic social prediction: Why deadlines put everyone under more pressure than everyone else. *Personality and Social Psychology Bulletin, 31*(5), 668–679.

Morrison, K. L., Madden, G. J., Odum, A. L., Friedel, J. E., & Twohig, M. P. (2014). Altering impulsive decision making with an acceptance-based procedure. *Behavior Therapy, 45*(5), 630–639.

Muir, W. M. (1985). Relative efficiency of selection for performance of birds housed in colony cages based on production in single bird cages. *Poultry Science, 64*(12), 2239–2247.

Muir, W. M. (1996). Group selection for adaptation to multiple-hen cages: Selection program and direct responses. *Poultry Science, 75*(4), 447–458.

Mutual Aid Networks. (n.d.). Mutual aid networks [Video, information on a page]. Retrieved from http://www.mutualaidnetwork.org

Neate, R. (2017, October 26). World's witnessing a new Gilded Age as billionaires' wealth swells to $6tn. *Guardian.* Retrieved from https://www.theguardian.com/business/2017/oct/26/worlds-witnessing-a-new-gilded-age-as-billionaires-wealth-swells-to-6tn

Nettle, D., Harper, Z., Kidson, A., Stone, R., Penton-Voak, I. S., & Bateson, M. (2013). The watching eyes effect in the Dictator Game: It's not how much you give, it's being seen to give something. *Evolution and Human Behavior, 34*(1), 35–40.

New Belgium Brewing. (2018). Business as a force for good. Retrieved from https://www.newbelgium.com/globalassets/sustainability/force-for-good-digital-report.pdf

Oc, B., Bashshur, M. R., & Moore, C. (2015). Speaking truth to power: The effect of candid feedback on how individuals with power allocate resources. *Journal of Applied Psychology, 100*(2), 450–463.

Ostrom, E. (2003). Toward a behavioral theory linking trust, reciprocity and reputation. In E. Ostrom & J. Walker (Eds.), *Trust and reciprocity: Interdisciplinary lessons for experimental research* (pp. 19–79). New York: Russell Sage Foundation.

Ostrom, E. (2008). *How do institutions for collective action evolve?* Delhi: Institute of Economic Growth.

Ostrom, E. (2010a). Beyond markets and states: Polycentric governance of complex economic systems. *American Economic Review, 100*(3), 641–672.

Ostrom, E. (2010b). Polycentric systems for coping with collective action and global environmental change. *Global Environmental Change, 20*(4), 550–557. doi:10.1016/j.gloenvcha.2010.07.004

Ostrom, E., & Parks, R. B. (1973). *Suburban police departments: Too many and too small?* Thousand Oaks, CA: Sage.

Parks, C. D., Joireman, J., & van Lange, P. A. (2013). Cooperation, trust, and antagonism: How public goods are promoted. *Psychological Science in the Public Interest, 14*(3), 119–165.

Participatory budgeting. (n.d.). In *Wikipedia*. Retrieved from https://en.wikipedia.org/wiki/Participatory_budgeting

Polk, K. L., Schoendorff, B., Webster, M., & Olaz, F. O. (2016). *The essential guide to the ACT matrix: A step-by-step approach to using the ACT matrix model in clinical practice*. Oakland, CA: Context Press.

P2P Foundation. (2017, October 27). Belonging to a superpower—patterns for decentralised organising. Retrieved from https://blog.p2pfoundation.net/belonging-is-a-superpower-patterns-for-decentralised-organising/2017/10/27

P2P Foundation. (n.d.). 1. Case study: Wikihouse. Retrieved from https://primer.commonstransition.org/4-more/5-elements/case-studies/case-study-wikihouse

P2P Foundation. (n.d.). 8. Case study: Sensorica. Retrieved from https://primer.commonstransition.org/4-more/5-elements/case-studies/case-study-sensorica

P2P Foundation. (n.d.). 9. Case study: Mutual aid networks. Retrieved from https://primer.commonstransition.org/4-more/5-elements/case-studies/case-study-mutual-aid-networks

P2P Foundation. (n.d.). 12. Case study: Bologna regulation for the urban commons. Retrieved from https://primer.commonstransition.org/4-more/5-elements/case-studies/case-study-bologna-regulation-for-the-care-and-regeneration-of-the-urban-commons

Raffone, A., & Srinivasan, N. (2017). Mindfulness and cognitive functions: Toward a unifying neurocognitive framework. *Mindfulness, 8*(1), 1–9.

Robertson, B. J. (2015). *Holacracy: The new management system for a rapidly changing world*. New York: Henry Holt.

Ryan, R. M., & Brown, K. W. (2003). Why we don't need self-esteem: On fundamental needs, contingent love, and mindfulness. *Psychological Inquiry, 14*(1), 71–76.

Ryan, R. M., & Deci, E. L. (2017). *Self-determination theory: Basic psychological needs in motivation, development, and wellness*. New York: Guilford Press.

Sawaoka, T., Hughes, B. L., & Ambady, N. (2015). Power heightens sensitivity to unfairness against the self. *Personality and Social Psychology Bulletin, 41*(8), 1023–1035.

Saxegaard, P. L. (2018, July 14). Humanizing corporations: A Nobel Prize for enlightened business leaders. *Evonomics*. Retrieved from http://evonomics.com/humanizing-corporations-a-nobel-prize-for-enlightened-business-leaders

Schäfer, M., Haun, D. B., & Tomasello, M. (2015). Fair is not fair everywhere. *Psychological Science, 26*(8), 1252–1260.

Schaffer, R. H., & Ashkenas, R. N. (2005). *Rapid results! How 100-day projects build the capacity for large-scale change*. New York: Jossey-Bass.

Schneider, S. M. (2012). *The science of consequences: How they affect genes, change the brain, and impact our world.* New York: Prometheus Books.

Seeley, T. D. (2010). *Honeybee democracy.* Princeton, NJ: Princeton University Press.

Shahzad, K., & Muller, A. R. (2016). An integrative conceptualization of organizational compassion and organizational justice: A sensemaking perspective. *Business Ethics: A European Review, 25*(2), 144–158.

Sheldon, K. M., Sheldon, M. S., & Osbaldiston, R. (2000). Prosocial values and group assortation: Within an N-person prisoner's dilemma. *Human Nature, 11*(4), 387–404.

Simons, T. L., & Peterson, R. S. (2000). Task conflict and relationship conflict in top management teams: The pivotal role of intragroup trust. *Journal of Applied Psychology, 85*(1), 102–111.

Skarlicki, D. P., & Folger, R. (1997). Retaliation in the workplace: The roles of distributive, procedural, and interactional justice. *Journal of Applied Psychology, 82*(3), 434–443.

Skinner, B. F. (1973). *Beyond freedom and dignity.* Harmondsworth, UK: Penguin.

Slemp, G. R., Kern, M. L., Patrick, K. J., & Ryan, R. M. (2018). Leader autonomy support in the workplace: A meta-analytic review. *Motivation and Emotion, 42*(5), 706–724.

Stewart, G. L., Courtright, S. H., & Barrick, M. R. (2012). Peer-based control in self-managing teams: Linking rational and normative influence with individual and group performance. *Journal of Applied Psychology, 97*(2), 435–447.

Stouten, J., & Tripp, T. M. (2009). Claiming more than equality: Should leaders ask for forgiveness? *Leadership Quarterly, 20*(3), 287–298.

Tekleab, A. G., Quigley, N. R., & Tesluk, P. E. (2009). A longitudinal study of team conflict, conflict management, cohesion, and team effectiveness. *Group & Organization Management, 34*(2), 170–205.

Travers, J., & Milgram, S. (1969). An experimental study of the small-world problem. *Social Networks, 32*, 425–443.

Turchin, P. (2016). *Ultrasociety: How 10,000 years of war made humans the greatest cooperators on Earth.* Chaplin, CT: Beresta Books.

Turillo, C. J., Folger, R., Lavelle, J. J., Umphress, E. E., & Gee, J. O. (2002). Is virtue its own reward? Self-sacrificial decisions for the sake of fairness. *Organizational Behavior and Human Decision Processes, 89*(1), 839–865.

Tyler, T. R., & Blader, S. L. (2013). *Cooperation in groups: Procedural justice, social identity, and behavioral engagement*: New York: Routledge Press.

Tyler, T. R., & Degoey, P. (1995). Collective restraint in social dilemmas: Procedural justice and social identification effects on support for authorities. *Journal of Personality and Social Psychology, 69*(3), 482–497.

Ury, W. (2015). *Getting to yes with yourself (and other worthy opponents).* New York: HarperCollins.

Useem, J. (2017, July/August). Power causes brain damage. *Atlantic.* Retrieved from https://www.theatlantic.com/magazine/archive/2017/07/power-causes-brain-damage/528711/

Van Bunderen, L., Greer, L. L., & van Knippenberg, D. (2017). When interteam conflict spirals into intrateam power struggles: The pivotal role of team power structures. *Academy of Management Journal, 61*(3), 1100–1130.

Van den Bos, K., & Lind, E. A. (2002). Uncertainty management by means of fairness judgments. *Advances in Experimental Social Psychology, 34*, 1–60.

Van den Bos, W., van Dijk, E., & Crone, E. A. (2011). Learning whom to trust in repeated social interactions: A developmental perspective. *Group Processes & Intergroup Relations, 15*(2), 243–256.

Van Lange, P. A. M., Joireman, J., Parks, C. D., & van Dijk, E. (2013). The psychology of social dilemmas: A review. *Organizational Behavior and Human Decision Processes, 120*(2), 125–141.

Van Vugt, M., & Hart, C. M. (2004). Social identity as social glue: The origins of group loyalty. *Journal of Personality and Social Psychology, 86*(4), 585–598.

Van Vugt, M., Jepson, S. F., Hart, C. M., & de Cremer, D. (2004). Autocratic leadership in social dilemmas: A threat to group stability. *Journal of Experimental Psychology, 40*(1), 1–13.

Varra, A. A., Hayes, S. C., Roget, N., & Fisher, G. (2008). A randomized control trial examining the effect of acceptance and commitment training on clinician willingness to use evidence-based pharmacotherapy. *Journal of Consulting and Clinical Psychology, 76*(3), 449–458.

Veage, S., Ciarrochi, J., Deane, F. P., Andresen, R., Oades, L. G., & Crowe, T. P. (2014). Value congruence, importance and success and in the workplace: Links with well-being and burnout amongst mental health practitioners. *Journal of Contextual Behavioral Science, 3*(4), 258–264.

Vilardaga, R., Estévez, A., Levin, M. E., & Hayes, S. C. (2012). Deictic relational responding, empathy, and experiential avoidance as predictors of social anhedonia: Further contributions from relational frame theory. *Psychological Record, 62*(3), 409–432.

Villatte, M., Villatte, J. L. & Hayes, S. C. (2015). *Mastering the clinical conversation: Language as intervention*. New York: Guilford Press.

Walton, G. M., & Cohen, G. L. (2011). A brief social-belonging intervention improves academic and health outcomes of minority students. *Science, 331*(6023), 1447–1451.

Walumbwa, F. O., Cropanzano, R., & Hartnell, C. A. (2009). Organizational justice, voluntary learning behavior, and job performance: A test of the mediating effects of identification and leader-member exchange. *Journal of Organizational Behavior, 30*(8), 1103–1126.

Wenzlaff, E. M., & Wegner, D. M. (2000). Thought suppression. *Annual Review of Psychology, 51*, 59–91.

Williams, M. J. (2014). Serving the self from the seat of power: Goals and threats predict leaders' self-interested behavior. *Journal of Management, 40*(5), 1365–1395.

Wilson, D. S. (2007). *Evolution for everyone: How Darwin's theory can change the way we think about our lives*. New York: Delacorte.

Wilson, D. S. (2015). *Does altruism exist? Culture, genes, and the welfare of others*. New Haven, CT: Yale University Press.

Wilson, D. S. (2017, June 2). Diagnosing a high-end application of PROSOCIAL: A conversation with Robert Styles. Prosocial Magazine. Retrieved from https://www.prosocial.world/post/diagnosing-a-high-end-application-of-prosocial-a-conversation-with-robert-styles

Wilson, D. S. (2019). *This view of life: Completing the Darwinian revolution*. New York: Pantheon Books.

Wilson, D. S., & Hessen, D. O. (2014). Blueprint for the global village. *Cliodynamics, 5*(1), 123–157.

Wilson, D. S., Kauffman Jr., R. A., & Purdy, M. S. (2011). A program for at-risk high school students informed by evolutionary science. *PLoS One, 6*(11), e27826.

Wilson, D. S., Marshall, D., & Iserhott, H. (2011). Empowering groups that enable play. *American Journal of Play, 3*(4), 523–537.

Wilson, D. S., & Wilson, E. O. (2007). Rethinking the theoretical foundation of sociobiology. *Quarterly Review of Biology, 82*(4), 327–348.

Yang, F., Choi, Y.-J., Misch, A., Yang, X., & Dunham, Y. (2018). In defense of the commons: Young children negatively evaluate and sanction free riders. *Psychological Science, 29*(10), 1598–1611.

Yong, E. (2016). *I contain multitudes: The microbes within us and a grander view of life*. New York: Random House.

Zettle, R. D., Hayes, S. C., Barnes-Holmes, D., & Biglan, A. (2016). *The Wiley handbook of contextual behavioral science*. Chichester, UK: John Wiley and Sons.

Zhen, S., & Yu, R. (2016). Tend to compare and tend to be fair: The relationship between social comparison sensitivity and justice sensitivity. *PLoS One, 11*(5), 1–17.

About the authors

Paul W.B. Atkins, PhD, is director of the Prosocial Institute, and senior research fellow with the Institute for Positive Psychology and Education at Australian Catholic University (ACU). His career has combined research and practice helping groups to thrive. His research has been published in the world's leading management and psychology journals. As a facilitator and coach, he has helped thousands of managers and leaders to improve their team leadership, conflict resolution, situational awareness, and well-being.

David Sloan Wilson, PhD, is president of The Evolution Institute, and SUNY distinguished professor of biology and anthropology at Binghamton University. He applies evolutionary theory to all aspects of humanity, in addition to the biological world. His books include *Darwin's Cathedral*, *Evolution for Everyone*, *The Neighborhood Project*, and *Does Altruism Exist?*

Steven C. Hayes, PhD, is Nevada Foundation Professor in the behavior analysis program in the department of psychology at the University of Nevada, Reno. His career has focused on analysis of the nature of human language and cognition, and its application to the understanding and alleviation of human suffering and promotion of human prosperity. Among other associations, Hayes has been president of the Association for Behavioral and Cognitive Therapies (ABCT), and the Association for Contextual Behavioral Science (ACBS). His work has received several awards, including the Lifetime Achievement Award from ABCT.

Foreword writer **Richard M. Ryan, PhD**, is a clinical psychologist and codeveloper of self-determination theory (SDT), an internationally recognized leading theory of human motivation. He is professor at the Institute for Positive Psychology and Education at ACU.

Index

A

about this book, 5–7
acceptance and commitment therapy (ACT): boot camps for training in, 207; peer-reviewed trainers community for, 135; psychological flexibility in, 56
ACT matrix. *See* matrix
action: committed, 62–63; long-term view of, 65
advice process, 143
after action reviews, 133
agile process, 156
allocation according to need, 121
altruism, 4, 13, 33, 45
ancient learning processes, 50–53
antecedents of behavior, 52
Anthropocene age, 2
antisocial behaviors, 4, 163
applied evolutionary science, 220
assessment, measurements for, 95
Association for Contextual Behavioral Science (ACBS), 135
associative learning, 55
attentional flexibility, 63–64
authority figures, 138, 139
authority to self-govern, 42–43, 189–198
autocratic decision making, 142
autonomy, viii, 42, 168, 178, 190
auxiliary design principles, 45
avoidance, experiential, 61–62
awareness, 63–64
away/toward dimension, 74–75

B

B Corps, 205
Bartlett, Rich, 108
beehives, 19
behavior change: ancient learning processes and, 50–53; challenges related to, 49–50; committed action and, 62–63; evolutionary principles of, 50–56; long-term thinking and, 65; psychological flexibility and, 56–69; social value orientation and, 65–67; symbolic learning and, 53–56; trust building and, 64–65
behaviors: encouraging cooperative, 165–166; monitoring agreed-upon, 40, 151–161; responding to, 40–41, 167–175
benefits, equitable distribution of, 38–39
between-group competition, 199, 204, 220
between-group selection, 12–13
Bockarie, Hannah, 102
Bollier, David, 30, 33
bottle-feeding infants, 25
bottom-left matrix quadrant, 78, 80, 127–129, 180
bottom-right matrix quadrant, 78, 79–80, 92, 125–126
boundaries, clearly defined, 24
brainstorming: digital tools for, 87; goal setting through, 212–213
bureaucracies, 145, 192
business: prosociality in, 204–205; rapid results in, 193–194
Business for Peace Foundation, 205
Buurtzorg system, 192–193, 202

C

call centers, 155, 167
Cambrian Period, 51–52
Camus, Albert, 206
Canberra Innovation Network, 44, 207, 211
cancerous cells, 13–14
care labor, 132
Carlin, George, 223
CDPs. *See* core design principles

centralized planning, 28–29
change: measurements for evaluating, 96. *See also* behavior change
cheating, decreased, 153
chicken breeding experiment, 14–15
circle talk, 148
classical conditioning, 51–52
coercive behavior, 138, 139, 155, 161
cognitive defusion, 62
cognitive flexibility, 62
cohesion in groups, 102–107; building shared identity and, 104–107; description of shared identity and, 102–104
Colbert, Stephen, 223
collaborative relations between groups, 43–44, 199–208
collective choice arrangements, 24
collective interests: mapping with collective matrix, 109–112; value of shared purpose and, 109
collective matrix, 73, 94, 96; conflict resolution and, 179, 188; decision making based on, 140–141; improving fairness using, 124–133; mapping collective interests with, 109–112; monitoring behavior using, 159–160; responding to behavior using, 168–171, 172; self-governing authority and, 196–197
college student experiment, 16
command-and-control decision making, 139, 140
committed action, 62–63
Common Cause study, 27
commoning: decision making and, 148–149; process of, 30, 33
common-pool resources, 23–24, 30
commons: building of, 218–222; derivation of word, 30; modern examples of, 30–33; parable of the tragedy of, 22–23, 26
communication skills, 106, 131, 184
community gardens, 32
competition: between-group, 199, 204, 220; narrative of, 206–207; within-group, 166, 202
conflict resolution, 24, 41–42, 176–188; fear and judgments preventing, 180; interpersonal skills for, 181–184; levels of effective, 181–187; matrix used for, 179, 182–184, 188; perspective taking and, 185–187; psychological flexibility and, 184–185; skills and processes for, 177, 181–187; strategies for avoiding, 180–181; task vs. relationship conflict and, 177–178
conflict-escalation agreement, 187
conformity, risk of, 106
consensus decision making, 143
Consensus-Oriented Decision-Making (Hartnett), 142
consent-based decision making, 143
consequences of behavior, 52
consultative leadership, 142
contingencies, 52, 55
contributions, equitable, 38–39
contributors vs. members, 105
conversations: focusing about fairness, 120–124; Prosocial guide to difficult, 184
cooperation: encouraging behaviors indicative of, 165–167; human distinctiveness and, 19–20; inclusive decision making and, 137; Internet related to, 20, 29, 30; problems of group, 4–5, 44; psychological flexibility and, 64–67; safe space created for, 163–165; shared purpose and, 108; social value orientation and, 67; symbolic learning and, 56
coordinating behavior, 153–154, 161
core design principles (CDPs), 24, 35–48; auxiliary design principles and, 45; group implementation of, 25–26, 46, 47; as multilevel framework, 44–45; Ostrom vs. Prosocial versions of, 35–36; practice exercise on, 47–48; psychological flexibility and, 222, 224;

scale independence of, 200–201; summary overview of, 37–44
core design principles (specific), 37–44; principle 1: shared identity and purpose, 37–38, 102–116; principle 2: equitable distribution of contributions and benefits, 38–39, 117–134; principle 3: fair and inclusive decision making, 39, 135–150; principle 4: monitoring agreed behaviors, 40, 151–161; principle 5: graduated responding to helpful and unhelpful behavior, 40–41, 162–175; principle 6: fast and fair conflict resolution, 41–42, 176–188; principle 7: authority to self-govern, 42–43, 189–198; principle 8: collaborative relations with other groups, 43–44, 199–208
counterproductive behaviors, 164
Covey, Stephen, 35
creative hopelessness, 85
Cronin, Katherine, 137
cultural evolution, 17–18, 221, 225
culture: human distinctiveness and, 19–20; inclusive decision making and, 144
cynicism, 223, 224

D

Darwin, Charles, 10–12
de Blok, Jos, 202
Deci, Ed, 165
decision making, 39, 135–150; benefits of inclusive, 136–138; collective matrix for, 140–141; culture and inner work of, 144; hierarchies and, 138–140, 145–148; integrating into groups, 140–148; models of, 142–144; power dynamics and, 139, 148, 149; procedural fairness and, 133, 149; question of commoning in, 148–149
defensive routines, 129–130
delay discounting, 65
delegation processes, 145–146
design principles. *See* core design principles

diagnosis, measurements for, 95
digital brainstorming tools, 87
dispassionate curiosity, 94
distributive fairness, 120, 122, 124, 133, 134
Dutch health care system, 192–193

E

Ebert, Beate, 135
Ebola crisis, 1, 102, 163
economic man, 26
ecosystem, 204, 207
egg production experiment, 14
Eisenstein, Charles, 218
emergency situations, 193, 194
emotions: evaluation of, 61; managing in groups, 93–94; reactive, 173
empathy, 131, 174
empty chair technique, 187
End of Leadership, The (Kellerman), 149
engagement: building motivation and, 165–167; inclusive decision making and, 137; sanctions and diminishment of, 164
Enspiral initiative, 32, 105
epigenetic evolution, 17
equality approach, 121, 122
equitable distribution of contributions and benefits, 38–39, 117–134. *See also* fairness
equity approach, 121–122
evaluation: of change, 96; of emotions, 61; of performance, 157–158
evolution: applied science of, 220; Darwin's theory of, 10–12; learning related to, 17, 51; major transitions in, 18–20; medical model vs. perspective of, 203; multilevel selection theory of, 12–16; streams of inheritance in, 16–18
evolutionary biology, 221
exercises: feeling "toward" and "away," 75; me and us, 69; mindful listening, 88–90; noticing "outside" and "inside," 77; picking a guide or hero, 68; spoke diagram for group, 47–48
experiential avoidance, 61–62

F

Fable of the Bees, The (Mandeville), 28
facilitative leadership, 142
Facilitator's Guide to Participatory Decision Making (Kaner), 142
fair and inclusive decision making, 39
fairness, 38–39, 117–134; distributive, 120, 122, 124, 133, 134; focusing conversations about, 120–124; hooks that work against, 127–129; importance of, 117, 119–120; improving with collective matrix, 124–133; increasing perceptions of, 131–133; preliminary ideas on improving, 126–127; procedural, 120, 133, 149; research institute example of, 118–119; values related to, 125–126
fast and fair conflict resolution, 41–42
federalism, political concept of, 190
feeling "toward" and "away" exercise, 75
flexible attention, 63–64
formal mindfulness practices, 88
"Frank case" role play, 174
Frank, Robert, 27
Freedman, Milton, 206
freedom, definition of, 74
Freeman, Jo, 148
free-riding behavior, 153
functional analytic psychotherapy (FAP), 158
functional contextual view, 221
functional design principles, 191–195

G

genetics, 16–17
Gintis, Herbert, 206
goals: characteristics of good, 210–212; matrix for clarifying, 145; minicase on working with, 214–216; performance vs. learning, 209–210; review process created for, 213–214; setting for committed action, 96; SMART acronym for, 212–213; values distinguished from, 59–60
gossip, 163–164
governance, polycentric, 24, 28–29, 33, 200–201, 204
graduated responding to helpful and unhelpful behavior, 40–41
graduated sanctions, 24, 162, 163, 167
Greene, Joshua, 4
Greenspan, Alan, 206
groups: clarifying membership in, 105; cohesion in, 102–107; collaborative relations between, 43–44, 199–208; cooperation problems of, 4–5, 44; core design principles for, 24, 35–48; decision making in, 135–150; facilitating the individual matrix with, 85–88; perceived fairness in, 117–134; polycentric governance of, 24, 28–29, 33, 200–201; self-governing authority of, 42–43, 189–198; shared identity and purpose in, 102–116; symbolic learning in, 55–56; values-based actions in, 62. *See also* small groups

H

habituation, 51
Harari, Yuval Noah, 26
Hardin, Garrett, 22–23, 30
Hartnett, Tim, 142
health care systems, 192–193
healthy vs. unhealthy conflict, 178
helpful behaviors: encouraging cooperative or, 165–166, 171; responding to, 40–41, 168–175. *See also* cooperation
hierarchies: decision making in traditional, 145–148; reasons for persistence of, 138–140
Holacracy approach, 132
Homo economicus, 26, 27, 29, 33, 66, 201, 206, 218
hope vs. cynicism, 224
hopelessness, creative, 85
"How Can We Make Conversations More Prosocial" guide, 184
human distinctiveness, 19–20

I

I/here/now awareness, 63
identity, shared. *See* shared identity
"Imagine" (Lennon), 223
impulsivity, reduction of, 65
inclusive decision making: benefits of, 136–138; culture and inner work of, 144; integrating into groups, 140–148
individual matrix, 73, 94, 104. *See also* collective matrix
individualistic response, 66
individuals, nature of, 56–57
ineffective monitoring, 154–155
inefficient sanctions, 165
inheritance, 16–18
insect colonies, 19
inside/outside dimension, 75–78
internal conversation, 90
international aid, 194
Internet: commoning examples on, 30–32; cooperation related to, 20, 29; information sharing and, 146, 149
interpersonal pathway, 98–99
interpersonal skills, 181–184
invisible hand metaphor, 28

J

Jackall, Robert, 202
Jentz, Barry, 174
Jobs, Steve, 114
Johansson, Magnus, 92
judgments, defusing from, 173

K

Kaner, Sam, 142, 144
Kellerman, Barbara, 149
King, Martin Luther, Jr., 224

L

laddering approach, 113
laissez-faire approach, 28, 29
language, symbolic learning and, 53–55
leadership: consultative, 142; facilitative, 142; hierarchical, 138–140
learning: ancient processes of, 50–53; behavior change and, 50–56; inheritance and, 17; symbolic processes of, 53–56, 59; written forms of, 55
learning circles, 31–32
learning goals, 209–210
Lemon, Jim, 203, 207
Lennon, John, 223
Lincoln, Abraham, 185
Linux operating system, 32
listening, mindful, 88–90
long-term thinking, 65
loyal dissent, 147

M

Mandeville, Bernard, 28
Markham, Edwin, 176
matrix, 72–94; clarifying goals and values using, 145; conflict resolution using, 179, 182–184; dimensions of experience on, 74–78; exercises on using, 75, 77–78; four quadrants of, 78, 79–83; group facilitation of, 85–94; individual vs. collective, 73; monitoring related to, 159–160; process of completing, 78–84; reinforcing belonging through, 104; shifting perspectives using, 186–187; targeting psychological flexibility in, 88–90; tips and tricks for facilitating, 90–94; workability of top-left behavior, 84–85. *See also* collective matrix
McChrystal, Stanley, 151, 194
me and us exercise, 69
"me versus us" problem, 4, 182, 199
meaning, contingencies of, 55
measurable goals, 211
measurement tools: for assessment and diagnosis, 95; for evaluating change, 96
mechanistic view, 220, 221
medical model, 203
meetings: monitoring through, 156–157; strategies for improving, 146–148
members vs. contributors, 105

memories, painful, 61
metastasis, 13
Milgram, Stanley, 3
military strategy, 151, 194–195
mindfulness, 63–64; formal practices of, 88; listening exercise based on, 88–90; monitoring related to, 158
mirroring process, 139
MLS theory, 12–16, 56, 69, 199
modules in Prosocial process, 95–96
monitoring, 24, 40, 151–161; benefits of, 153–154; collective matrix and, 159–160; positive use of, 156–158; psychological flexibility and, 158–159; varieties of, 151–153; wrong use of, 154–155
Moral Mazes: The World of Corporate Managers (Jackall), 202
Moral Tribes (Greene), 4
morality: group behavior and, 11–12; prosociality and, 5
More Beautiful World Our Hearts Know Is Possible, The (Eisenstein), 218
motivation: building engagement and, 165–167; monitoring related to, 153; shared purpose and, 108, 114
Muir, William, 14
multigroup organizations, 202–204
multigroup populations, 12
multilevel selection (MLS) theory, 12–16, 56, 69, 199
Museum of Australian Democracy, 1–2
mutual accountability, 157–158
mutual aid networks, 31

N

narrative of competition, 206–207
natural selection, 10–12
Navy SEAL teams, 195
needs: allocation according to, 121; meeting to encourage cooperation, 166
Neff, Kristen, 88
New Belgium Brewing Company, 204–205
niche construction, 18

noticing "outside" and "inside" exercise, 77, 88

O

observing self, 185
Olympic Games, 219–220
On the Origin of Species (Darwin), 11
onboarding processes, 106
open-access journals, 30–31
open-book management, 146
operant learning, 52–53, 54
optimal scale, 28
organizations, multigroup, 202–204
Ostrom, Elinor, vii, 3, 21, 22, 23–26, 28, 50, 190, 191, 200, 219
Ostrom, Vincent, 22, 28, 200
outside/inside dimension, 75–78
"Outwitted" (Markham), 176

P

painful memories, 61
pair-discussion approach, 85–86
pathways in Prosocial process, 97–101
peer-to-peer contracts, 108
performance goals, 209
performance management, 157–158
perspective broadening, 58
perspective taking, 63; conflict resolution and, 185–187; perceived fairness and, 131; responding based on, 174; shared purpose and, 112–113; shifting using the matrix, 186–187; tools for flexible, 186; trust related to, 65
picking a guide or hero exercise, 68
Polanyi, Michael, 201
polycentric governance, 24, 28–29, 33, 200–201, 204
power dynamics, 108, 139, 148, 149
privatization, 23
problem solving, 54, 55, 61
procedural fairness, 120, 133, 149
project-management tools, 157
prosocial behavior, 4–5, 13, 153, 161

Prosocial process: ACT used in, 56; behavior change and, 50–56; boot camps for training in, 207; core design principles in, 35–48; ecosystem level for, 207; evaluation tools for, 214; future of, 225; matrix used in, 72–94; modules and pathways for, 95–101; psychological flexibility in, 56–69; social value orientation and, 67; three main fronts of, 3–4; trust building and, 64–65
prosocial.world website, 47, 97
prosociality, 5, 67, 153, 204–205
psychological flexibility, 50, 56–69; awareness and, 63–64; conflict resolution and, 184–185; cooperation and, 64–67; core design principles and, 222, 224; decision making and, 144; description of, 56–59; exercises for exploring, 68–69; goal setting related to, 210; long-term thinking and, 65; monitoring and, 158–159; perceived fairness and, 132; retention and, 62–63; social value orientation and, 65–67; targeting in the matrix, 88–90; trust and, 64–65; values and, 59–60; variation and, 60–62
purpose, shared. *See* shared purpose

R

Rand, Ayn, 28
rapid results cycles, 193–194
Rapid Results Institute, 194
rational choice theory, 28
reflective listening, 88
Regents Academy, 2
relational learning, 53–56
relationships: conflicts related to, 178; power dynamics in, 108, 139, 148, 149
relevant goals, 211–212
reparation ceremony, 102
representative democracy, 148, 149
reputational rating system, 152
resilient systems, 137
responding to behavior: based on core design principles, 167–168; steps for effectively, 168–175
responsibility, viii, 42, 108, 137
retention, 11, 62–63
reward systems, 165–166
roles, focusing on, 132
Ryan, Richard, ix, 165

S

sanctions, 163–164, 168
Sapiens (Harari), 26
scale independence, 200–201
Schaffer, Robert H., 193
Seeley, Thomas, 19
selection, 11, 12–13
self-determination, viii, 136, 137, 161
self-governing authority, 42–43, 189–198
self-interested behaviors, 4, 138–139
Sensorica network, 32
shared identity, 37–38, 102–107; building group cohesion and, 104–107; description of group cohesion and, 102–104
shared purpose, 37–38, 107–115; collective matrix for building, 109–112; creating a strong sense of, 107; getting the level right for, 113–114; importance of living, 114–115; laddering approach to finding, 113; perspective taking related to, 112–113; practical benefits of, 107–109; reorienting individuals to, 171, 173
similarity, creation of, 106–107
six degrees of separation, 3
Skinner, B. F., 17, 74
Slemp, Gavin, 136
small groups: evolutionary changes and, 3; polycentric governance and, 28–29; utilizing the power of, 104–105. *See also* groups
"small world" experiment, 3
SMART goals, 212–213
Smith, Adam, 28
social cohesion, 103
social dilemma, 32, 50

social interactions, 166
social learning, 52
social protocols, 30
social value orientation, 65–67
software organizations, 156
speaking up in groups, 146–147
spoke diagram exercise, 47–48
stakeholder perspectives, 146
standard pathway, 97–98
strategic planning pathway, 99–100
Styles, Robert, 191, 214
subsidiarity, concept of, 190
surveys, 47, 133
symbolic learning, 53–56, 59
symbols: cooperation related to, 20; learning related to, 53–56

T

tasks: cohesion related to, 103; conflicts related to, 177–178; focusing on roles and, 132
Team of Teams (McChrystal), 151, 153, 194
Thatcher, Margaret, 206
thinking, long-term, 65
time-bound goals, 212
Tocqueville, Alexis de, 3
tomato-processing company, 108, 114
top-down regulation, 20, 23
top-left matrix quadrant, 78, 81–82, 84–85, 129–130
top-right matrix quadrant, 78, 82–83, 126–127, 131
toward/away dimension, 74–75
tragedy of the commons, 22–23, 26
transgressions, responding to, 40–41
transparency: behavioral, 151, 152; information, 146
trial-and-error learning, 17

trust: cooperation and sense of, 64–65; fairness related to, 120; group cohesion and, 105; matrix use related to, 91–92; sanctions and diminishment of, 164
"Tyranny of Structurelessness, The" (Freeman), 148

U

uncooperative behaviors: discouraging, 163–165; intentional vs. unintentional, 4; responding to, 171–175
unfairness, perceptions of, 117
unhealthy vs. healthy conflict, 178
unhelpful behaviors, responding to, 40–41, 168–175
"us versus them" problem, 4, 182, 199

V

values, 59–60; discovering and clarifying, 60, 61; fairness-related, 125–126; goals distinguished from, 59; group actions based on, 62; matrix for clarifying, 79–80, 145; sharing of, 65; social, 67
variation, 11, 60–62
verbal problem solving, 54
vulnerability, 91

W

Walras, Léon, 28
well-being, 137–138
whole-group approach, 86–87
Wikihouse, 32
Wikipedia, 30
Wilson, Edward O., 13
within-group competition, 166
within-group selection, 13
word clouds, 87
written language, 55

MORE BOOKS *from* NEW HARBINGER PUBLICATIONS

ACT MADE SIMPLE, SECOND EDITION
An Easy-To-Read Primer on Acceptance & Commitment Therapy
978-1684033010 / US $44.95

MINDFULNESS & ACCEPTANCE WORKBOOK FOR STRESS REDUCTION
Using Acceptance & Commitment Therapy to Manage Stress, Build Resilience & Create the Life You Want
978-1684031283 / US $24.95

MESSAGES, FOURTH EDITION
The Communications Skills Book
978-1684031719 / US $21.95

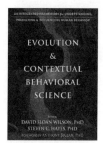

EVOLUTION & CONTEXTUAL BEHAVIORAL SCIENCE
An Integrated Framework for Understanding, Predicting & Influencing Human Behavior
978-1626259133 / US $39.95
CONTEXT PRESS
An Imprint of New Harbinger Publications

THE MINDFUL & EFFECTIVE EMPLOYEE
An Acceptance & Commitment Therapy Training Manual for Improving Well-Being & Performance
978-1608820214 / US $49.95

IT'S TIME TO TALK (AND LISTEN)
How to Have Constructive Conversations About Race, Class, Sexuality, Ability & Gender in a Polarized World
978-1684032679 / US $16.95

newharbingerpublications
1-800-748-6273 / newharbinger.com

(VISA, MC, AMEX / prices subject to change without notice)

Follow Us

Don't miss out on new books in the subjects that interest you.
Sign up for our Book Alerts at **newharbinger.com/bookalerts**

A subsidiary of New Harbinger Publications, Inc.

Enhance your practice with live ACT workshops

Praxis Continuing Education and Training—a subsidiary of New Harbinger Publications—is the premier provider of evidence-based continuing education for mental health professionals. Praxis specializes in ongoing **acceptance and commitment therapy (ACT)** training—taught by leading ACT experts. Praxis workshops are designed to help professionals learn and effectively implement ACT in session with clients.

- **ACT BootCamp®: Introduction to Implementation**
  For professionals with no prior experience with ACT, as well as those who want to refresh their knowledge

- **ACT 1: Introduction to ACT**
  For professionals with no prior experience with ACT

- **ACT 2: Clinical Skills-Building Intensive**
  For professionals who practice ACT, but want more hands-on experience

- **ACT 3: Mastering ACT**
  For professionals actively using ACT who want to apply it to their most complex cases

Receive 10% off Any Training!

Visit praxiscet.com and use code **EDU10** at checkout to receive 10% off.

Check out our workshops and register now at praxiscet.com

Can't make a workshop? Check out our online and on-demand courses at **praxiscet.com**

CONCEPTUAL | EXPERIENTIAL | PRACTICAL